OXFORD CHEMISTRY PRIMERS

1 S. E. Thomas *Organic synthesis: The roles of boron and silicon*
2 D. T. Davies *Aromatic heterocyclic chemistry*
3 P. R. Jenkins *Organometallic reagents in synthesis*
4 M. Sainsbury *Aromatic chemistry*
5 L. M. Harwood *Polar rearrangements*
7 J. H. Jones *Amino acid and peptide synthesis*
8 C. J. Moody and G. H. Whitham *Reactive intermediates*
9 G. M. Hornby and J. M. Peach *Foundations of organic chemistry*
10 R. Henderson *The mechanisms of reactions at transition metal sites*
11 H. M. Cartwright *Applications of artificial intelligence in chemistry*
12 M. Bochmann *Organometallics 1: Complexes with transition metal–carbon s-bonds*
13 M. Bochmann *Organometallics 2: Complexes with transition metal–carbon p-bonds*
14 C. E. Housecroft *Cluster molecules of the p-block elements*
15 M. J. Winter *Chemical bonding*
16 N. C. Norman *Periodicity and the p-block elements*
17 R. S. Ward *Bifunctional compounds*
18 S. K. Scott *Oscillations, waves, and chaos in chemical kinetics*
19 T. P. Softley *Atomic spectra*
20 J. Mann *Chemical aspects of biosynthesis*
21 B. G. Cox *Modern liquid phase kinetics*
22 A. Harrison *Fractals in chemistry*
23 M. T. Weller *Inorganic materials chemistry*
24 R. P. Wayne *Chemical instrumentation*
25 D. E. Fenton *Biocoordination chemistry*
26 W. G. Richards and P. R. Scott *Energy levels in atoms and molecules*
27 M. J. Winter *d-Block chemistry*
28 D. M. P. Mingos *Essentials of inorganic chemistry 1*
29 G. H. Grant and W. G. Richards *Computational chemistry*
30 S. A. Lee and G. E. Robinson *Process development: Fine chemicals from grams to kilograms*
31 C. L. Willis and M. R. Wills *Organic synthesis*
32 P. J. Hore *Nuclear magnetic resonance*
33 G. H. Whitham *Organosulfur chemistry*
34 A. C. Fisher *Electrode dynamics*
35 G. D. Meakins *Functional groups: Characteristics and interconversions*
36 A. J. Kirby *Stereoelectronic effects*
37 P. A. Cox *Introduction to quantum theory and atomic structure*
38 P. D. Bailey and K. M. Morgan *Organonitrogen chemistry*
39 C. E. Wayne and R. P. Wayne *Photochemistry*
40 C. P. Lawrence, A. Rodger, and R. G. Compton *Foundations of physical chemistry*
41 R. G. Compton and G. H. W. Sanders *Electrode potentials*
42 P. B. Whalley *Two-phase flow and heat transfer*
43 L. M. Harwood and T. D. W. Claridge *Introduction to organic spectroscopy*
... *l bonded carbonyl dimers and*
45 H. Maskill *Mechanisms of organic reactions*
46 P. C. Wilkins and R. G. Wilkins *Inorganic chemistry in biology*
47 J. H. Jones *Core carbonyl chemistry*
48 N. J. B. Green *Quantum mechanics 1: Foundations*
49 I. S. Metcalfe *Chemical reaction engineering: A first course*
50 R. H. S. Winterton *Heat transfer*
51 N. C. Norman *Periodicity and the s- and p-block elements*
52 R. W. Cattrall *Chemical sensors*
53 M. Bowker *The Basis and applications of heterogeneous catalysis*
54 M. C. Grossel *Alicyclic chemistry*
55 J. M. Brown *Molecular spectroscopy*
56 G. J. Price *Thermodynamics of chemical processes*
57 A. G. Howard *Aquatic environmental chemistry*
58 A. O. S. Maczek *Statistical thermodynamics*
59 G. A. Attard and C. J. Barnes *Surfaces*
60 W. Clegg *Crystal structure determination*
61 M. Brouard *Reaction dynamics*
62 A. K. Brisdon *Inorganic spectroscopic methods*
63 G. Proctor *Stereoselectivity in organic synthesis*
64 C. M. A. Brett and A. M. O. Brett *Electroanalysis*
65 N. J. B. Green *Quantum mechanics 2: The tool kit*
66 D. M. P. Mingos *Essentials of inorganic chemistry 2*
67 I. Fleming *Pericyclic reactions*
68 N. S. Lawrence, J. D. Wadhawan & R. G. Compton *Foundations of physical chemistry: Worked examples*
69 J. R. Chipperfield *Non-aqueous solvents*
70 T. J. Mason *Sonochemistry*
71 J. McCleverty *Chemistry of the first-row transition metals*
72 E. C. Constable *Coordination chemistry of macrocyclic compounds*
73 C. E. Housecroft *The heavier d-block metals: Aspects of inorganic and coordination chemistry*
74 P. D. Beer, P. A. Gale and D. K. Smith *Supramolecular chemistry*
75 A. F. Orchard *Magnetochemistry*
76 N. Kaltsoyannis and P. Scott *The f elements*
77 D. S. Sivia and S. G. Rawlings *Foundations of science mathematics*
78 S. Duckett and B. Gilbert *Foundations of spectroscopy*
79 J. H. Atherton and K. J. Carpenter *Process development: physicochemical concepts*
80 R. de Levie *Acid–base equilibria and titrations*
81 H. Maskill *Structure and reactivity in organic chemistry*
82 D. S. Sivia and S. G. Rawlings *Foundations of science mathematics: Worked problems*
83 J. Iggo *NMR Spectroscopy in inorganic chemistry*
84 P. Biggs *Computers in chemistry*
85 D. J. Walton and P. Lorimer *Polymers*
86 R. Davey and J. Garside *From molecules to crystallizers*
87 G. M. Hornby and J. M. Peach *Fundamentals of organic chemistry: Worked problems*

Foundations of Science Mathematics: Worked Problems

D. S. Sivia
Rutherford Appleton Laboratory and St John's College, Oxford

S. G. Rawlings
Department of Astrophysics and St Peter's College, Oxford

OXFORD
UNIVERSITY PRESS

This book has been printed digitally and produced in a standard specification in order to ensure its continuing availability

OXFORD
UNIVERSITY PRESS

Great Clarendon Street, Oxford OX2 6DP

Oxford University Press is a department of the University of Oxford.
It furthers the University's objective of excellence in research, scholarship, and education by publishing worldwide in

Oxford New York

Auckland Cape Town Dar es Salaam Hong Kong Karachi
Kuala Lumpur Madrid Melbourne Mexico City Nairobi
New Delhi Shanghai Taipei Toronto

With offices in

Argentina Austria Brazil Chile Czech Republic France Greece
Guatemala Hungary Italy Japan South Korea Poland Portugal
Singapore Switzerland Thailand Turkey Ukraine Vietnam

Oxford is a registered trade mark of Oxford University Press in the UK and in certain other countries

Published in the United States
by Oxford University Press Inc., New York

Reprinted 2007

ISBN 978-0-19-850429-0

Printed and bound by CPI Antony Rowe, Eastbourne

Series Editor's Foreword

Oxford Chemistry Primers are designed to provide clear and concise introductions to a wide range of topics that may be encountered by chemistry students as they progress from the freshman stage through to graduation. The Physical Chemistry series contains books easily recognised as relating to established fundamental core material that all chemists need to know, as well as books reflecting new directions and research trends in the subject, thereby anticipating (and perhaps encouraging) the evolution of modern undergraduate courses.

In this Physical Chemistry Primer Devinder Sivia and Steve Rawlings present a book of worked examples to complement their elegant introductory account of *Foundations of Science Mathematics* (Oxford Chemistry Primer 77). The two books together explain and illustrate in simple terms the basic ideas and applications of a subject which is essential knowledge for any practising scientist. These Primers will be of interest to all students of science (and their mentors).

Richard G. Compton
Physical and Theoretical Chemistry Laboratory,
University of Oxford

Preface

Mathematics plays a central role in the life of every scientist and engineer, from the earliest school days to late on into the college and professional years. While the subject may be met with enthusiasm or reluctance, depending on individual taste, all are required to learn it to some degree at university level. This Primer is the 'worked examples' half of a two-volume set that summarises the basic concepts and results which sould be familiar from high school, and then extends the ideas to cover the material needed by the majority of undergraduates. It assumes a familiarity with the theoretical content of the subject explained in the main maths Primer (OCP 77), and aims to bolster the reader's practical confidence by providing 'model answers', and additional comments, for a wide range of exercises.

We would like to thank several friends and colleagues who showed keen interest in the project, and helped to facilitate its speedy conclusion: namely, Drs. Jerry Mayers, Jeff Penfold and Andrew Taylor. Professor Richard Compton's guidance and encouragement was invaluable throughout.

Oxford D. S. S.
April, 1999 S. G. R.

Contents

1 Basic algebra and arithmetic

1.1 Evaluate: **(a)** $4^{3/2}$, **(b)** $27^{-2/3}$, **(c)** $3^2\,3^{-3/2}$, **(d)** $\log_2(8)$ and **(e)** $\log_2(8^3)$.

(a) $4^{3/2} = 4^{1/2\times 3} = \left(4^{1/2}\right)^3 = \left(\sqrt{4}\right)^3 = 2^3 = \underline{8}$

or $= 4^{1+1/2} = 4^1\,4^{1/2} = 4\sqrt{4} = 4\times 2 = \underline{8}$

(b) $27^{-2/3} = \dfrac{1}{27^{2/3}} = \dfrac{1}{\left(\sqrt[3]{27}\right)^2} = \dfrac{1}{3^2} = \underline{\dfrac{1}{9}}$

(c) $3^2\,3^{-3/2} = 3^{2-3/2} = 3^{1/2} = \underline{\sqrt{3}}$

(d) $\log_2(8) = \log_2(2^3) = \underline{3}$

(e) $\log_2(8^3) = 3\log_2(8) = 3\times 3 = \underline{9}$

or $= \log_2\left[(2^3)^3\right] = \log_2(2^9) = \underline{9}$

1.2 By letting $A = a^M$ and $B = a^N$, and using the definition of a logarithm, show that

$$\log(AB) = \log(A) + \log(B) \qquad \text{and} \qquad \log(A/B) = \log(A) - \log(B)$$

Similarly, show that

$$\log(A^\beta) = \beta\log(A) \qquad \text{and} \qquad \log_b(A) = \log_a(A)\times\log_b(a)$$

$$A = a^M \iff M = \log_a(A) \qquad \text{and} \qquad B = a^N \iff N = \log_a(B)$$

$$\text{But}\ \ AB = a^M a^N = a^{M+N}$$

$$\therefore\ \log_a(AB) = \log_a\left(a^{M+N}\right) = M+N = \log_a(A) + \log_a(B)$$

$$\text{i.e.}\ \ \underline{\log(AB) = \log(A) + \log(B)}$$

This result holds for logarithms to any base, because we did not specify the value of a above.

$$\frac{1}{B} = \frac{1}{a^N} = a^{-N}$$

$$\therefore \quad \log_a\left(\frac{1}{B}\right) = \log_a\left(a^{-N}\right) = -N = -\log_a(B)$$

Hence, using the previous result for the logarithm of the product of A and $1/B$, we obtain

$$\underline{\log(A/B) = \log(A) - \log(B)}$$

$$A^\beta = \left(a^M\right)^\beta = a^{M\beta}$$

$$\therefore \quad \log_a\left(A^\beta\right) = \log_a\left(a^{M\beta}\right) = M\beta = \log_a(A)\,\beta$$

$$\text{i.e.} \quad \underline{\log\left(A^\beta\right) = \beta\log(A)}$$

$$\log_b(A) = \log_b\left(a^M\right) = M\log_b(a)$$

$$\therefore \quad \underline{\log_b(A) = \log_a(A) \times \log_b(a)}$$

1.3 Derive the formula for the two solutions of the quadratic equation $ax^2 + bx + c = 0$.

$$\text{If } a \neq 0, \qquad x^2 + \frac{bx}{a} + \frac{c}{a} = 0$$

$$\therefore \quad \left(x + \frac{b}{2a}\right)^2 - \frac{b^2}{4a^2} + \frac{c}{a} = 0$$

$$\therefore \quad \left(x + \frac{b}{2a}\right)^2 = \frac{b^2 - 4ac}{4a^2}$$

$$\therefore \quad x + \frac{b}{2a} = \frac{\pm\sqrt{b^2 - 4ac}}{2a}$$

$$\text{i.e.} \quad \underline{x = \frac{-b \pm \sqrt{b^2 - 4ac}}{2a}}$$

While this formula is only valid if $a \neq 0$, the case of $a = 0$ is even simpler because we are then left with the linear equation $bx + c = 0$; this has the solution $x = -c/b$.

1.4 Solve: **(a)** $x^2-5x+6=0$, **(b)** $3x^2+5x-2=0$ and **(c)** $x^2-4x+2=0$.

(a) $x^2-5x+6=(x-3)(x-2)=0 \qquad \therefore\ \underline{x=2 \text{ or } x=3}$

(b) $3x^2+5x-2=(3x-1)(x+2)=0 \qquad \therefore\ \underline{x=-2 \text{ or } x=1/3}$

If this factorisation is difficult to spot, then the result could be obtained from the general formula of exercise 1.3.

$$x=\frac{-5\pm\sqrt{25+24}}{6}=\frac{-5\pm7}{6}=-\tfrac{12}{6} \text{ or } \tfrac{2}{6}$$

(c) $x=\frac{4\pm\sqrt{16-8}}{2}=2\pm\frac{\sqrt{8}}{2}=2\pm\sqrt{\frac{8}{4}} \qquad$ i.e. $\underline{x=2\pm\sqrt{2}}$

1.5 For what values of k does $x^2+kx+4=0$ have real roots?

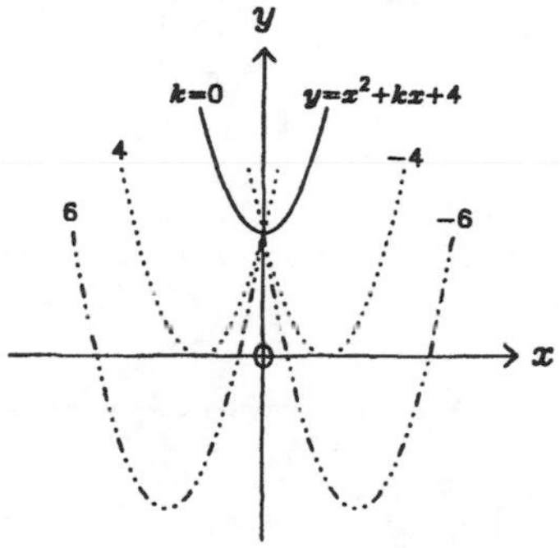

For real roots, "$b^2\geq 4ac$". Therefore, we require that $k^2\geq 16$.

$$\therefore\ \underline{|k|\geq 4}$$

In other words, we need k to be less than -4 or greater than $+4$.

1.6 Solve the following simultaneous equations:

(a) $\begin{aligned}3x+2y&=4\\ x-7y&=9\end{aligned}$, **(b)** $\begin{aligned}x^2+y^2&=2\\ x-2y&=1\end{aligned}$, **(c)** $\begin{aligned}3x+2y+5z&=0\\ x+4y-2z&=9\\ 4x-6y+3z&=3\end{aligned}$

(a) $3x+2y=4 \quad$ — (1)

$x-7y=9 \quad$ — (2)

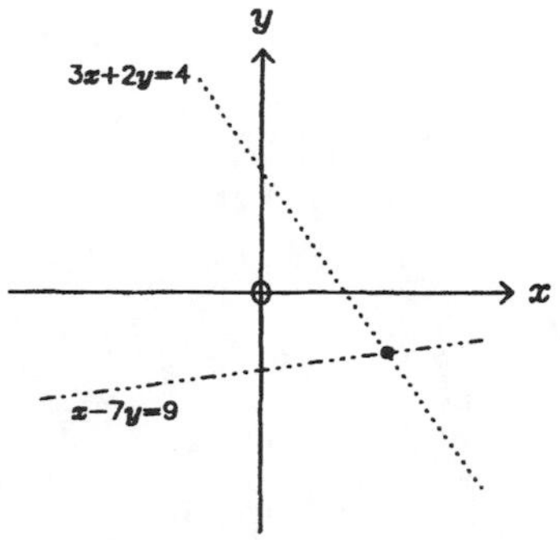

$(1)-3\times(2) \Rightarrow \quad 3x+2y-(3x-21y)=4-27$

$\therefore \quad 23y=-23$

$\therefore \quad y=-1, \text{ and } (2)\Rightarrow x+7=9$

i.e. $\underline{x=2 \text{ and } y=-1}$

(b) $x^2+y^2=2 \quad — (3)$

$x-2y=1 \quad — (4)$

Substituting $x=2y+1$ from (4) into (3) :

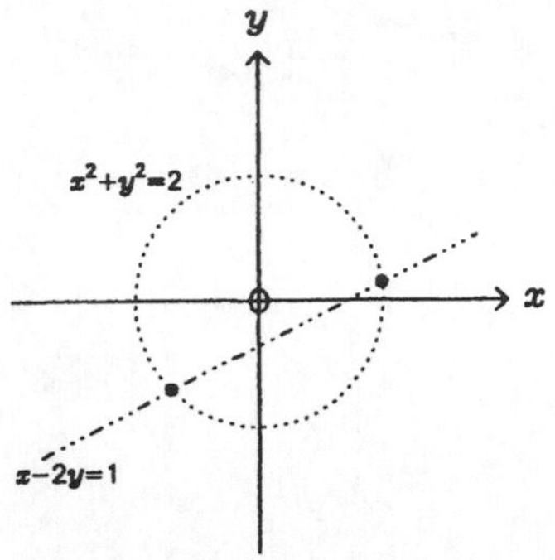

$$(2y+1)^2+y^2=2$$

$$\therefore\quad 4y^2+4y+1+y^2=2$$

$$\therefore\quad 5y^2+4y-1=0$$

$$\therefore\quad (5y-1)(y+1)=0$$

$$\therefore\quad y=\tfrac{1}{5} \quad \text{or} \quad y=-1$$

When $y=1/5$, (4) $\Rightarrow$ $x=1+2/5$

When $y=-1$, (4) $\Rightarrow$ $x=1-2$

Hence $\underline{x=7/5 \ \text{and}\ y=1/5, \ \text{or}\ x=-1 \ \text{and}\ y=-1}$

(c) $3x+2y+5z=0 \quad — (5)$

$x+4y-2z=9 \quad — (6)$

$4x-6y+3z=3 \quad — (7)$

$$\left.\begin{aligned}(5)-3\times(6) &\Rightarrow -10y+11z=-27\\ (7)-4\times(6) &\Rightarrow -22y+11z=-33\end{aligned}\right\} \quad \begin{aligned}y&=1/2\\ z&=-2\end{aligned}$$

$$\therefore\quad \underline{x=3,\ y=1/2\ \text{and}\ z=-2}$$

1.7 Derive the formulae for the sums of an arithmetic progression and a geometric progression.

Let $a+(a+d)+(a+2d)+\cdots+(l-2d)+(l-d)+l=S_N \quad — (1)$

where $l=a+(N-1)d$

$$\therefore\quad l+(l-d)+(l-2d)+\cdots+(a+2d)+(a+d)+a=S_N \quad — (2)$$

$$(1)+(2) \Rightarrow (a+l)+(a+l)+\cdots+(a+l)+(a+l)=2S_N$$

$$\therefore \quad 2S_N = N(a+l) = N\big[2a+(N-1)d\big]$$

$$\text{i.e.} \quad \underline{\text{Sum of AP} = \tfrac{N}{2}\big[2a+(N-1)d\big]}$$

$$\text{Let} \qquad a+ar+ar^2+\cdots+ar^{N-2}+ar^{N-1} = S_N \qquad \text{— (3)}$$

$$r\times(3) \Rightarrow \qquad ar+ar^2+\cdots+ar^{N-2}+ar^{N-1}+ar^N = rS_N \qquad \text{— (4)}$$

$$(4)-(3) \Rightarrow \qquad a(r^N-1) = S_N(r-1)$$

$$\text{i.e.} \quad \underline{\text{Sum of GP} = \frac{a(r^N-1)}{r-1}}$$

> **1.8** By expressing a recurring decimal number as the sum of an infinite GP, show that 0.121212··· = 4/33. What is 0.3181818··· as a fraction?

$$0.12121212\cdots = 0.12+0.0012+0.000012+\cdots$$

$$= \text{sum of infinite GP with } a = 0.12 \text{ and } r = 0.01$$

$$= \frac{0.12}{1-0.01}$$

$$= \frac{12}{99}$$

$$\therefore \quad \underline{0.12121212\cdots = 4/33}$$

$$0.318181818\cdots = 0.3+0.018+0.00018+0.0000018+\cdots$$

$$= \frac{3}{10}+\frac{0.018}{1-0.01}$$

$$= \frac{3}{10}+\frac{18}{990}$$

$$= \frac{33}{110}+\frac{2}{110}$$

$$= \frac{35}{110}$$

$$\therefore \quad \underline{0.318181818\cdots = 7/22}$$

1.9 Decompose the following into partial fractions:

(a) $\dfrac{1}{x^2-5x+6}$, **(b)** $\dfrac{x^2-5x+1}{(x-1)^2(2x-3)}$, **(c)** $\dfrac{11x+1}{(x-1)(x^2-3x-2)}$

(a) $$\frac{1}{x^2-5x+6}=\frac{1}{(x-3)(x-2)}=\frac{A}{(x-3)}+\frac{B}{(x-2)}$$

$$\therefore\ A(x-2)+B(x-3)=1$$

Putting $x=2 \Rightarrow B=-1$, and $x=3 \Rightarrow A=1$

i.e. $$\underline{\frac{1}{x^2-5x+6}=\frac{1}{(x-3)}-\frac{1}{(x-2)}}$$

We could, in fact, have written this answer straight down by using the "cover-up" rule.

(b) $$\frac{x^2-5x+1}{(x-1)^2(2x-3)}=\frac{A}{(x-1)^2}+\frac{B}{(x-1)}+\frac{C}{(2x-3)}$$

$$\therefore\ A(2x-3)+(x-1)\left[B(2x-3)+C(x-1)\right]=x^2-5x+1$$

Putting $x=1 \Rightarrow A=3$, and $x=3/2 \Rightarrow C=-17$

Equating coefficients of $x^2 \Rightarrow 2B+C=1 \quad \therefore\ B=9$

i.e. $$\underline{\frac{x^2-5x+1}{(x-1)^2(2x-3)}=\frac{3}{(x-1)^2}+\frac{9}{(x-1)}-\frac{17}{(2x-3)}}$$

In this question, A and C are readily obtained with the "cover-up" rule; B, however, requires the formal analysis.

(c) $$\frac{11x+1}{(x-1)(x^2-3x-2)}=\frac{A}{(x-1)}+\frac{Bx+C}{(x^2-3x-2)}$$

$$\therefore\ A(x^2-3x-2)+(x-1)(Bx+C)=11x+1$$

Putting $x=1 \Rightarrow A=-3$

Putting $x=0 \Rightarrow -2A-C=1 \quad \therefore\ C=5$

Equating coefficients of $x^2 \Rightarrow A+B=0 \quad \therefore\ B=3$

$$\text{i.e.}\quad \frac{11x+1}{(x-1)(x^2-3x-2)} = \frac{3x+5}{(x^2-3x-2)} - \frac{3}{(x-1)}$$

Here only A follows from the "cover-up" rule; both B and C need the more formal analysis.

1.10 Evaluate the following summations:

$$\textbf{(a)}\ \sum_{n=1}^{\infty} \frac{1}{n(n+1)} \quad \text{and} \quad \textbf{(b)}\ \sum_{n=0}^{\infty} e^{-\beta(n+1/2)}$$

(a)

$$\frac{1}{n(n+1)} = \frac{1}{n} - \frac{1}{(n+1)}$$

$$\begin{aligned}\therefore\ \sum_{n=1}^{\infty} \frac{1}{n(n+1)} &= 1 + \tfrac{1}{2} + \tfrac{1}{3} + \tfrac{1}{4} + \tfrac{1}{5} + \cdots \\ &\quad - \tfrac{1}{2} - \tfrac{1}{3} - \tfrac{1}{4} - \tfrac{1}{5} - \cdots \\ &= \underline{1}\end{aligned}$$

This is a simple example of how partial fractions can be useful; they are used frequently in integration problems.

(b)

$$\begin{aligned}\sum_{n=0}^{\infty} e^{-\beta(n+1/2)} &= e^{-\beta/2} + e^{-3\beta/2} + e^{-5\beta/2} + e^{-7\beta/2} + \cdots \\ &= \text{sum of infinite GP with } a = e^{-\beta/2} \text{ and } r = e^{-\beta} \\ &= \underline{\frac{e^{-\beta/2}}{1-e^{-\beta}}}\end{aligned}$$

While the above example was merely an exercise in recognising and evaluating the sum of an infinite GP, it does relate to a problem of physical interest. In quantum mechanics, the solutions of the Schrödinger equation for a particle in a harmonic potential (such as that found for a diatomic molecule) show that the energy levels for the system are given by

$$E_n = \left(n + \tfrac{1}{2}\right) h\nu$$

where $n = 0, 1, 2, 3, \ldots$, h is *Planck's* constant, and ν is the natural frequency of the vibrations given by the curvature of the potential well; the case of $n = 0$ is known as the *ground state*, with a *zero-point* energy of $E_0 = h\nu/2$. The probability that a particle occupies an energy level E_n is given by the *Boltzmann* factor $\exp(-E_n/kT)$, where k is the Boltzmann constant and T is the temperature (in Kelvin). The sum of these occupation probabilities yields the summation above, with $\beta = h\nu/kT$, and is known as the *partition function*.

2 Curves and graphs

2.1 Find the equation of the straight line that passes through the two points $(-1,3)$ and $(3,1)$; where does it intersect with $y = x + 1$?

General equation of a straight line is $y = mx + c$.

$$\left.\begin{array}{lll}\text{Passes through } (-1,3) & \Rightarrow & 3 = -m + c \\ \text{Passes through } (3,1) & \Rightarrow & 1 = 3m + c\end{array}\right\} \quad \begin{array}{l} m = -1/2 \\ c = 5/2 \end{array}$$

$\therefore$ Equation of line is $\underline{2y = 5 - x}$

$$\text{For intersection,} \quad \left.\begin{array}{r} 2y = 5 - x \\ y = 1 + x \end{array}\right\} \quad \begin{array}{l} y = 2 \\ x = 1 \end{array}$$

$\therefore$ Lines intersect at $\underline{(1,2)}$

2.2 By 'completing the square', find the coordinates of the turning point of $y = x^2 + x + 1$; hence sketch the parabola.

$$y = x^2 + x + 1 = \left(x + \tfrac{1}{2}\right)^2 - \tfrac{1}{4} + 1$$

$$= \left(x + \tfrac{1}{2}\right)^2 + \tfrac{3}{4}$$

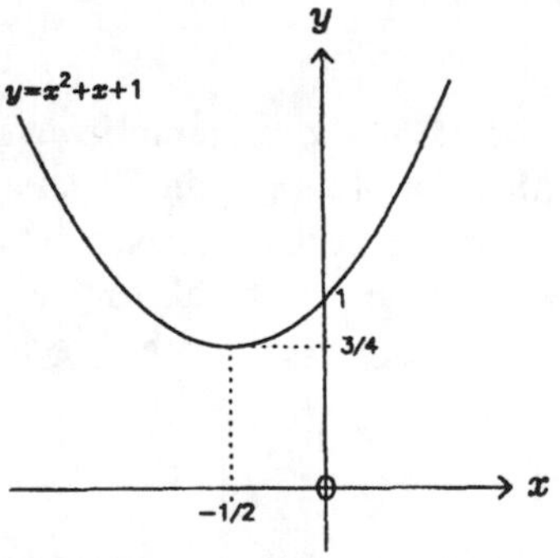

Smallest value of y when $x + 1/2 = 0$

$\therefore$ Turning point is at $\underline{(-1/2, 3/4)}$

2.3 Find the equation of the parabola that passes through the three points $(0,3)$, $(3,0)$, and $(5,8)$; what are the roots of the equation?

General equation of a parabola is $y = ax^2 + bx + c$.

$$\left.\begin{array}{lll} \text{Passes through } (0,3) & \Rightarrow & 3 = c \\ \text{Passes through } (3,0) & \Rightarrow & 0 = 9a+3b+c \\ \text{Passes through } (5,8) & \Rightarrow & 8 = 25a+5b+c \end{array}\right\} \quad \begin{array}{l} c = 3 \\ a = 1 \\ b = -4 \end{array}$$

$\therefore$ Equation of parabola is $\underline{y = x^2 - 4x + 3}$

For roots, $y = 0 \Rightarrow x^2 - 4x + 3 = (x-3)(x-1) = 0$

$\therefore$ Roots are $\underline{x = 1 \text{ and } x = 3}$

> **2.4** Where does the curve $y = (x-3)(x-1)(x+1)$ cross the x and y axes? Hence sketch this cubic function, and state the ranges of x values for which it is greater than zero.

$y = 0$ when $x = 3$, $x = 1$ and $x = -1$.

When $x = 0$, $y = 3$.

$y > 0$ when $\underline{|x| < 1 \text{ or } x > 3}$

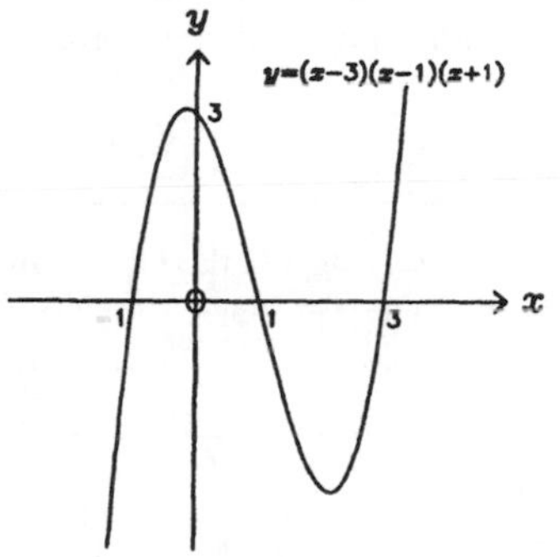

> **2.5** Sketch the functions $y = 1 - e^{-x}$ and $y = 1 - e^{-2x}$ for positive values of x; and $y = e^{-|x|}$, $y = 1/x$, and $y = 1/(x^2 - 1)$ for all x.

Since the exponential function e^{-x} decays from a value of one at $x = 0$ to zero as $x \to \infty$, $1 - e^{-x}$ rises from the origin to an asymptotic limit of $y = 1$ as x becomes very large; e^{-2x} decays twice as quickly as e^{-x}, therefore $1 - e^{-2x}$ rises faster than $1 - e^{-x}$ to its ultimate value of $y = 1$.

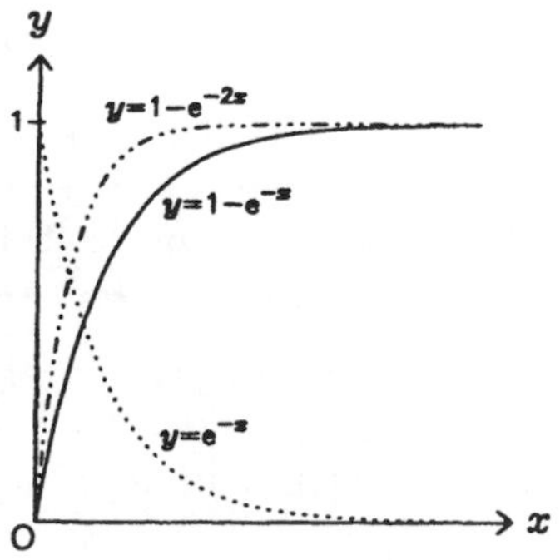

With reference to Chapter 4, neither $e^{-|x|}$ nor $1/x$ is differentiable at the origin. That is to say, because of their cuspy and discontinuous behaviour respectively, neither function has a well-defined gradient at $x = 0$. Also, $e^{-|x|}$ is an *even* or *symmetric* function and $1/x$ is an *odd* or *antisymmetric* function; this refers to the symmetry of the curve with respect to its reflection given by an imagined mirror along $x = 0$.

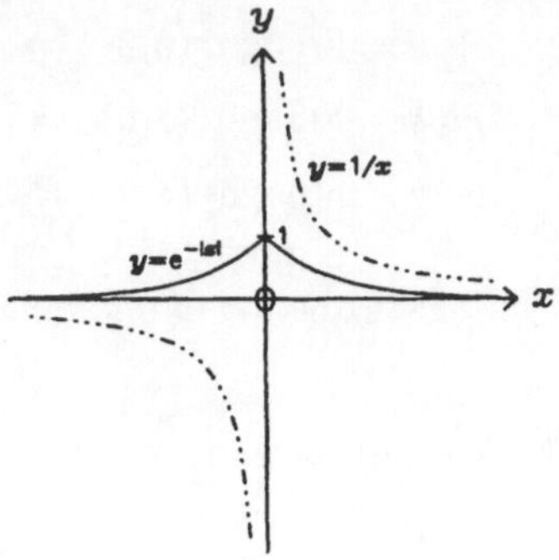

The best way of sketching a complicated function is often to decompose it into a series of more straightforward steps. For $1/(x^2 - 1)$, for example, first sketch the parabola $y = x^2 - 1$; then take its reciprocal, so that large values of y transform to smaller ones and vice versa.

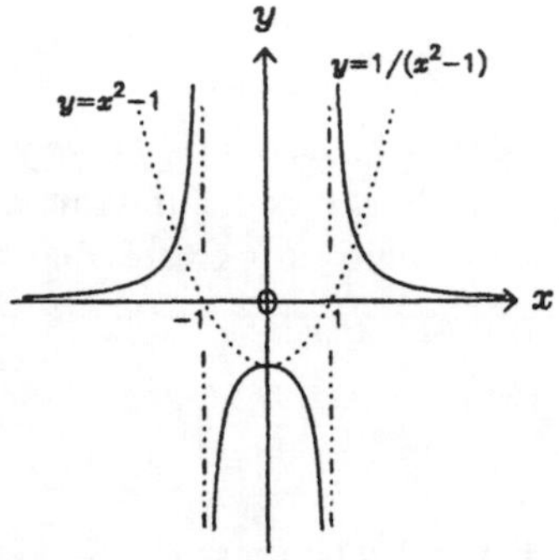

2.6 Given that the exponential dominates for large x, sketch the functions $y = xe^{-x}$ and $y = x^2e^{-x}$ for $x \geq 0$.

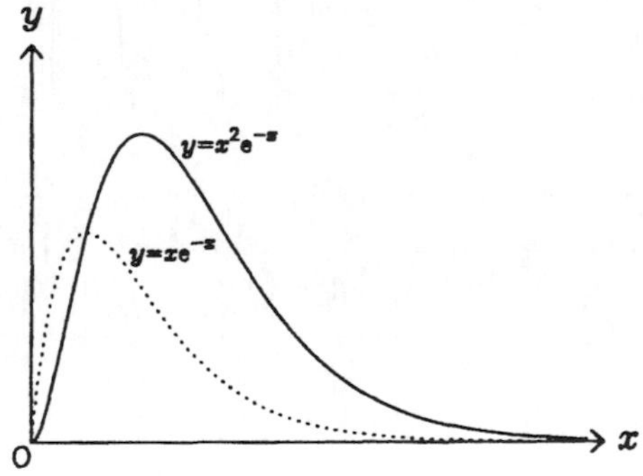

2.7 What is the centre and radius of the circle $x^2 + y^2 - 2x + 4y - 4 = 0$?

$$x^2 - 2x + y^2 + 4y = 4$$

$$\therefore \quad (x-1)^2 - 1 + (y+2)^2 - 4 = 4$$

$$\therefore \quad (x-1)^2 + (y+2)^2 = 9 = 3^2$$

i.e. centre is at $(1,-2)$ and radius $= 3$

This 'completing the square' transformation of the original equation into the form $(x-x_0)^2+(y-y_0)^2=r^2$ is useful because it then becomes obvious that we have a circle with centre (x_0,y_0) and radius r. Alternatively, we could remember that $x^2+y^2+2gx+2hy+c=0$ is a general equation for a circle; its centre is at $(-g,-h)$ and the radius is $\sqrt{g^2+h^2-c}$.

2.8 Sketch the ellipse $3x^2+4y^2=3$; evaluate its eccentricity, and indicate the positions of the focus and directrix.

The simplest equation of an ellipse is

$$\frac{x^2}{a^2}+\frac{y^2}{b^2}=1$$

where the principal axes are along the x and y axes, and of width $2a$ and $2b$ respectively. Thus, in our case,

$$x^2+\frac{4}{3}y^2=1$$

so that $a=1$ and $b=\sqrt{3}/2$.

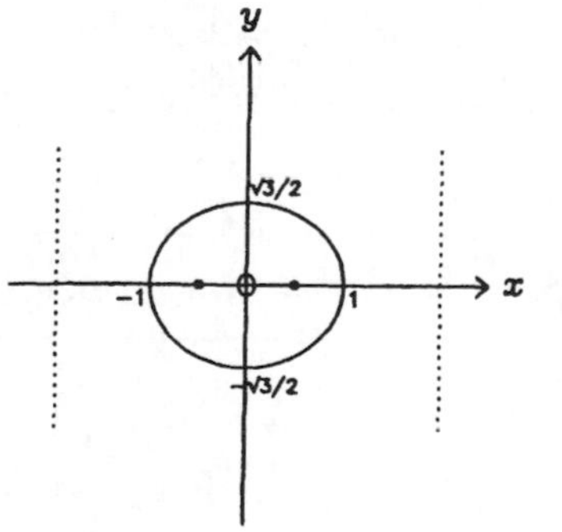

The eccentricity ε is related to a and b through $b^2=a^2(1-\varepsilon^2)$, giving $\varepsilon=1/2$. The foci are at $(\pm a\varepsilon,0)$, or $(\pm 1/2,0)$, and the two directrices are the straight lines $x=\pm a/\varepsilon=\pm 2$.

2.9 By first factorising the equation, or otherwise, sketch the function $(x^2+y^2)^2-4x^2=0$.

The expression is of the form a^2-b^2, and can be factorised as $(a-b)(a+b)$:

$$(x^2+y^2-2x)(x^2+y^2+2x)=0$$

$$\therefore \quad x^2 - 2x + y^2 = 0 \quad \text{or} \quad x^2 + 2x + y^2 = 0$$

$$\therefore \quad (x-1)^2 + y^2 = 1 \quad \text{or} \quad (x+1)^2 + y^2 = 1$$

i.e. We have two circles of radius 1, with centres at $(1,0)$ and $(-1,0)$.

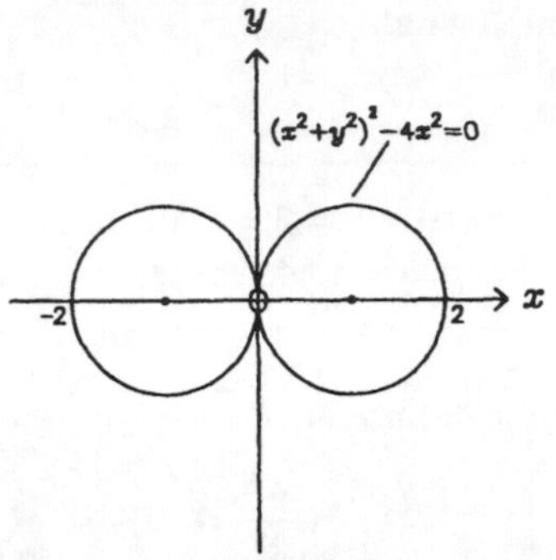

Incidentally, if the factorisation hint had not been given, or spotted, then an obvious way of reducing the complexity of the given equation would be to write it as

$$(x^2 + y^2)^2 = 4x^2$$

and take the square root of both sides. This gives

$$x^2 + y^2 = \pm 2x$$

which also leads to the equation of the two circles above.

2.10 Sketch the hyperbola $x^2 - y^2 = 1$, and mark in the asymptotes.

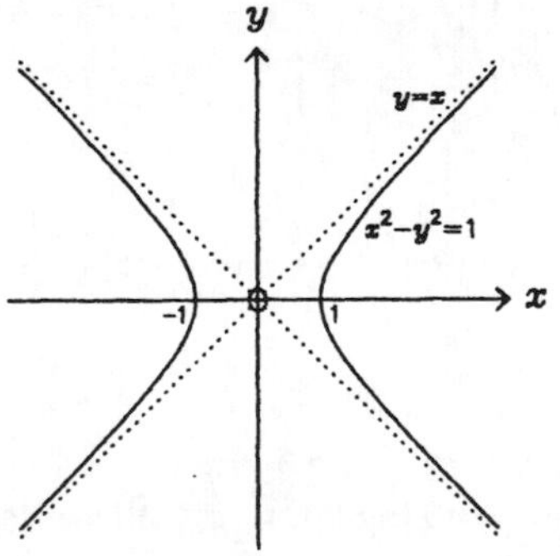

3 Trigonometry

3.1 From the definitions of a radian and a sine, indicate why $\sin\theta \approx \theta$ for small angles; show how this leads to $\cos\theta \approx 1-\theta^2/2$.

$$\sin\theta = \frac{AB}{OB} \quad \text{and} \quad \theta = \frac{\text{Arc-length } AC}{OC}$$

But as $\theta \to 0$, $\quad AB \to$ Arc-length AC

$\qquad\qquad OB \to OC$

$\therefore \quad \sin\theta \to \theta$

i.e. $\underline{\sin\theta \approx \theta}$ for $\theta \ll 1$

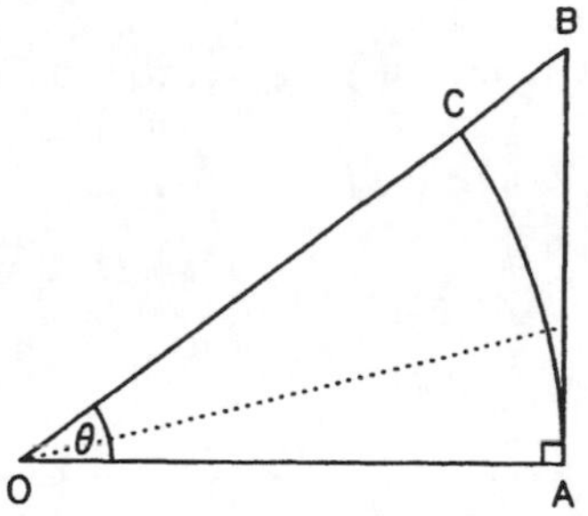

Also, "$\cos 2\phi = 1 - 2\sin^2\phi$"

$$\therefore \quad \cos\theta \approx 1 - 2\left(\frac{\theta}{2}\right)^2 \qquad \text{i.e.} \quad \underline{\cos\theta \approx 1 - \frac{\theta^2}{2}} \quad \text{for } \theta \ll 1$$

3.2 If $t = \tan(\theta/2)$, express $\tan\theta$, $\cos\theta$ and $\sin\theta$ in terms of t.

If $\tan(\theta/2) = t$,

Pythagoras' $\Rightarrow \quad \sin(\theta/2) = \dfrac{t}{\sqrt{1+t^2}} \quad$ and $\quad \cos(\theta/2) = \dfrac{1}{\sqrt{1+t^2}}$

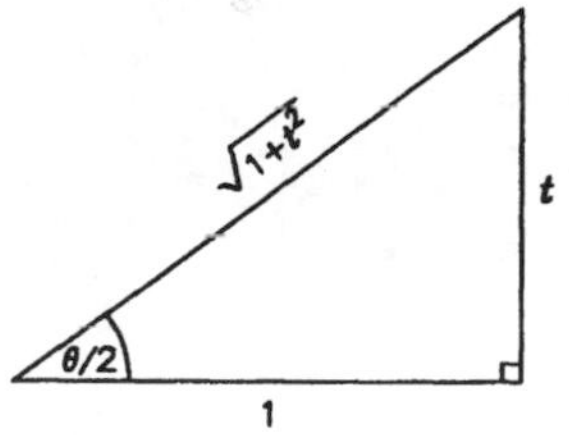

But $\sin\theta = 2\sin(\theta/2)\cos(\theta/2) \qquad \therefore \quad \underline{\sin\theta = \dfrac{2t}{1+t^2}}$

$\cos\theta = \cos^2(\theta/2) - \sin^2(\theta/2) \qquad \therefore \quad \underline{\cos\theta = \dfrac{1-t^2}{1+t^2}}$

$\tan\theta = \dfrac{\sin\theta}{\cos\theta} \qquad \therefore \quad \underline{\tan\theta = \dfrac{2t}{1-t^2}}$

This somewhat obscure substitution is occasionally useful for integrating trigonometric functions; its differential is given by

$$dt = \tfrac{1}{2}\sec^2(\theta/2)\,d\theta = \tfrac{1}{2}\left[1+\tan^2(\theta/2)\right]d\theta \qquad \text{i.e.} \quad \underline{d\theta = \frac{2\,dt}{1+t^2}}$$

3.3 Solve the following in the range $-\pi$ to π: **(a)** $\tan\theta = -\sqrt{3}$, **(b)** $\sin 3\theta = -1$, and **(c)** $4\cos^3\theta = \cos\theta$.

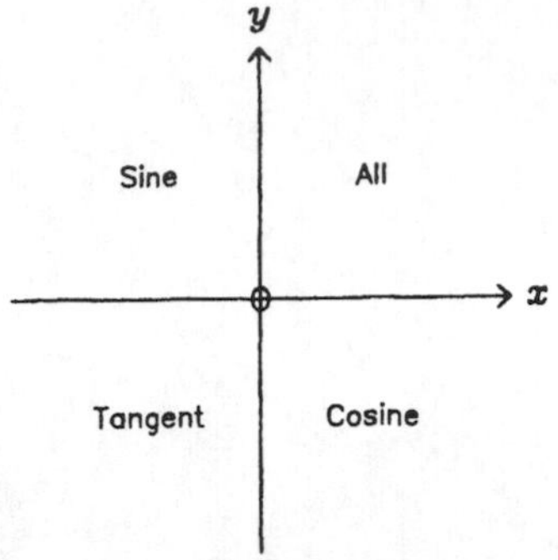

(a) $\tan\theta = -\sqrt{3} \;\Rightarrow\; \underline{\theta = -\frac{\pi}{3} \text{ or } \frac{2\pi}{3}}$

(b) $\sin 3\theta = -1 \;\Rightarrow\; 3\theta = -\frac{5\pi}{2} \text{ or } -\frac{\pi}{2} \text{ or } \frac{3\pi}{2}$

$$\therefore\quad \underline{\theta = -\frac{5\pi}{6} \text{ or } -\frac{\pi}{6} \text{ or } \frac{\pi}{2}}$$

(c) $4\cos^3\theta - \cos\theta = 0 \;\Rightarrow\; \cos\theta\,(4\cos^2\theta - 1) = 0$

$$\therefore\quad \cos\theta = 0 \quad\text{or}\quad \cos\theta = \pm\tfrac{1}{2}$$

$$\therefore\quad \underline{\theta = \pm\frac{\pi}{2} \quad\text{or}\quad \theta = \pm\frac{\pi}{3} \text{ or } \pm\frac{2\pi}{3}}$$

3.4 Show that $a\sin\theta + b\cos\theta$ can be written as $A\sin(\theta+\phi)$, where A and ϕ are related to a and b. Hence solve $\sin\theta + \cos\theta = \sqrt{3/2}$.

If $\quad a\sin\theta + b\cos\theta = A\sin(\theta+\phi)$

$$= A\cos\phi\sin\theta + A\sin\phi\cos\theta$$

Then $\left.\begin{aligned} A\cos\phi &= a \\ A\sin\phi &= b \end{aligned}\right\}\quad \begin{aligned} &\frac{\sin\phi}{\cos\phi} = \tan\phi = \frac{b}{a} \\ &a^2 + b^2 = A^2(\sin^2\phi + \cos^2\phi) = A^2 \end{aligned}$

$$\text{i.e.}\quad \underline{A = \sqrt{a^2+b^2} \quad\text{and}\quad \phi = \tan^{-1}(b/a)}$$

$$\sin\theta + \cos\theta = \sqrt{\tfrac{3}{2}} \;\Rightarrow\; \sqrt{2}\sin(\theta + \pi/4) = \sqrt{\tfrac{3}{2}}$$

$$\therefore\quad \sin(\theta+\pi/4) = \frac{\sqrt{3}}{2}$$

$$\therefore\quad \theta + \frac{\pi}{4} = \frac{\pi}{3} \text{ or } \frac{2\pi}{3}$$

$$\text{i.e. } \underline{\theta = \frac{\pi}{12} \text{ or } \frac{5\pi}{12}}$$

3.5 Show that $\cos 4\theta = 8\cos^4\theta - 8\cos^2\theta + 1$, and express $\sin 4\theta$ in terms of $\sin\theta$ and $\cos\theta$.

$$\cos 2\theta = 2\cos^2\theta - 1 \;\Rightarrow\; \cos 4\theta = 2\cos^2 2\theta - 1$$

$$= 2(2\cos^2\theta - 1)^2 - 1$$

$$= 2(4\cos^4\theta - 4\cos^2\theta + 1) - 1$$

$$= \underline{8\cos^4\theta - 8\cos^2\theta + 1}$$

$$\sin 2\theta = 2\sin\theta\cos\theta \;\Rightarrow\; \sin 4\theta = 2\sin 2\theta\cos 2\theta$$

$$= 2(2\sin\theta\cos\theta)(\cos^2\theta - \sin^2\theta)$$

$$= \underline{4\sin\theta\cos^3\theta - 4\sin^3\theta\cos\theta}$$

3.6 Show that $8\sin^4\theta = \cos 4\theta - 4\cos 2\theta + 3$, and find a similar expression for $\cos^4\theta$.

$$\cos 2\theta = 1 - 2\sin^2\theta = 2\cos^2\theta - 1$$

$$\therefore\; 8\sin^4\theta = 8(\sin^2\theta)^2 = 8\left(\frac{1-\cos 2\theta}{2}\right)^2$$

$$= 2(1 - 2\cos 2\theta + \cos^2 2\theta)$$

$$= 2 - 4\cos 2\theta + (2\cos^2 2\theta - 1) + 1$$

$$= \underline{3 - 4\cos 2\theta + \cos 4\theta}$$

$$\therefore\; 8\cos^4\theta = 8(\cos^2\theta)^2 = 8\left(\frac{\cos 2\theta + 1}{2}\right)^2$$

$$= 2(\cos^2 2\theta + 2\cos 2\theta + 1)$$

$$= (2\cos^2 2\theta - 1) + 1 + 4\cos 2\theta + 2$$

$$= \underline{\cos 4\theta + 4\cos 2\theta + 3}$$

3.7 Show that $\cos\theta + \cos 3\theta + \cos 5\theta + \cos 7\theta = 4\cos\theta\cos 2\theta\cos 4\theta$.

$$\cos A + \cos B = 2\cos\left(\tfrac{A+B}{2}\right)\cos\left(\tfrac{A-B}{2}\right)$$

$\cos(-\theta) = \cos\theta$

$$\therefore\quad \cos\theta + \cos 3\theta = 2\cos 2\theta\cos\theta \quad\text{and}\quad \cos 5\theta + \cos 7\theta = 2\cos 6\theta\cos\theta$$

$$\begin{aligned}\therefore\quad \cos\theta + \cos 3\theta + \cos 5\theta + \cos 7\theta &= 2\cos\theta(\cos 2\theta + \cos 6\theta)\\ &= 2\cos\theta(2\cos 4\theta\cos 2\theta)\\ &= \underline{4\cos\theta\cos 2\theta\cos 4\theta}\end{aligned}$$

3.8 By using a factor formula, find the values of θ between 0 and π which satisfy the equation $\cos\theta = \cos 2\theta + \cos 4\theta$.

$$\cos 2\theta + \cos 4\theta = 2\cos 3\theta\cos\theta$$

$$\therefore\quad \cos\theta = \cos 2\theta + \cos 4\theta \quad\Rightarrow\quad \cos\theta(1 - 2\cos 3\theta) = 0$$

$$\therefore\quad \cos\theta = 0 \quad\text{or}\quad \cos 3\theta = \tfrac{1}{2}$$

$$\therefore\quad \theta = \tfrac{\pi}{2} \quad\text{or}\quad 3\theta = \tfrac{\pi}{3} \text{ or } \tfrac{5\pi}{3} \text{ or } \tfrac{7\pi}{3}$$

$$\text{i.e. } \underline{\theta = \tfrac{\pi}{9} \text{ or } \tfrac{\pi}{2} \text{ or } \tfrac{5\pi}{9} \text{ or } \tfrac{7\pi}{9}}$$

3.9 A triatomic molecule has bond-lengths of 1.327Å and 1.514Å, and a bond-angle of 107.5°; find the distance between the furthest atoms.

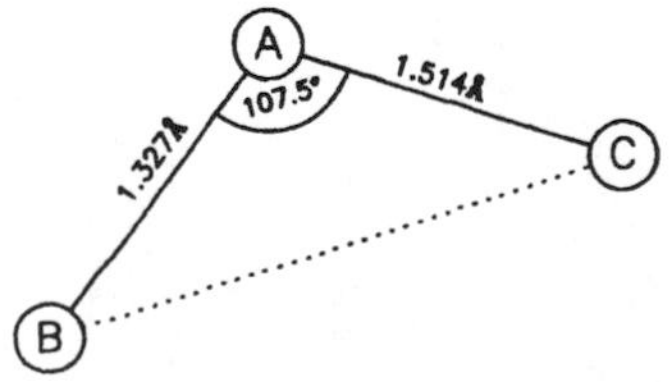

"Cosine rule": $BC^2 = AB^2 + AC^2 - 2\,AB.AC\cos(B\hat{A}C)$

$$\begin{aligned}\therefore\quad \text{Distance}^2 &= 1.327^2 + 1.514^2 - 2 \times 1.327 \times 1.514 \times \cos(107.5°)\\ &= 5.2614\ \text{Å}^2\end{aligned}$$

$\therefore$ Distance between furthest atoms $=$ $\underline{2.294\ \text{Å}}$

4 Differentiation

4.1 From 'first principles', differentiate: **(a)** $y = \cos x$; **(b)** $y = x^n$, where n is a positive integer; **(c)** $y = 1/x$; and **(d)** $y = 1/x^2$.

(a)
$$\frac{dy}{dx} = \lim_{\delta x \to 0}\left(\frac{\cos(x+\delta x) - \cos x}{\delta x}\right)$$
$$= \lim_{\delta x \to 0}\left(\frac{\cos(\delta x)\cos x - \sin(\delta x)\sin x - \cos x}{\delta x}\right)$$
$$= \lim_{\delta x \to 0}\left(\frac{(1 - \delta x^2/2)\cos x - \delta x\sin x - \cos x}{\delta x}\right)$$
$$= \lim_{\delta x \to 0}\left(-\frac{\delta x}{2}\cos x - \sin x\right)$$
$$\therefore \quad \frac{d}{dx}(\cos x) = -\sin x$$

(b)
$$\frac{dy}{dx} = \lim_{\delta x \to 0}\left(\frac{(x+\delta x)^n - x^n}{\delta x}\right)$$
$$= \lim_{\delta x \to 0}\left(\frac{x^n + nx^{n-1}\delta x + n(n-1)/2\, x^{n-2}\delta x^2 + \cdots - x^n}{\delta x}\right)$$
$$= \lim_{\delta x \to 0}\left(nx^{n-1} + \frac{n}{2}(n-1)x^{n-2}\delta x + \cdots\right)$$
$$\therefore \quad \frac{d}{dx}(x^n) = nx^{n-1}$$

(c)
$$\frac{dy}{dx} = \lim_{\delta x \to 0}\left(\frac{1/(x+\delta x) - 1/x}{\delta x}\right)$$
$$= \lim_{\delta x \to 0}\left(\frac{x - (x+\delta x)}{x(x+\delta x)\,\delta x}\right)$$
$$= \lim_{\delta x \to 0}\left(\frac{-1}{x(x+\delta x)}\right)$$

$$\therefore \quad \frac{\mathrm{d}}{\mathrm{d}x}\left(\frac{1}{x}\right) = -\frac{1}{x^2}$$

(d)
$$\frac{\mathrm{d}y}{\mathrm{d}x} = \lim_{\delta x\to 0}\left(\frac{1/(x+\delta x)^2 - 1/x^2}{\delta x}\right)$$

$$= \lim_{\delta x\to 0}\left(\frac{x^2-(x+\delta x)^2}{x^2\,(x+\delta x)^2\,\delta x}\right)$$

$$= \lim_{\delta x\to 0}\left(\frac{x^2-x^2-2x\delta x-\delta x^2}{x^2\,(x+\delta x)^2\,\delta x}\right)$$

$$= \lim_{\delta x\to 0}\left(\frac{-2}{x\,(x+\delta x)^2} - \frac{\delta x}{x^2\,(x+\delta x)^2}\right)$$

$$\therefore \quad \frac{\mathrm{d}}{\mathrm{d}x}\left(\frac{1}{x^2}\right) = -\frac{2}{x^3}$$

4.2 Give an argument for why the gradient of a straight line perpendicular to $y = mx + c$ is $-1/m$.

$$\text{Slope of } \overrightarrow{AB} = \tan\theta = \frac{BC}{AB}$$

$$\text{Slope of } \overrightarrow{BC} = -\tan\phi = -\frac{AB}{BC}$$

$$\therefore \quad \text{Slope of } \overrightarrow{AB} = \frac{-1}{\text{Slope of } \overrightarrow{BC}}$$

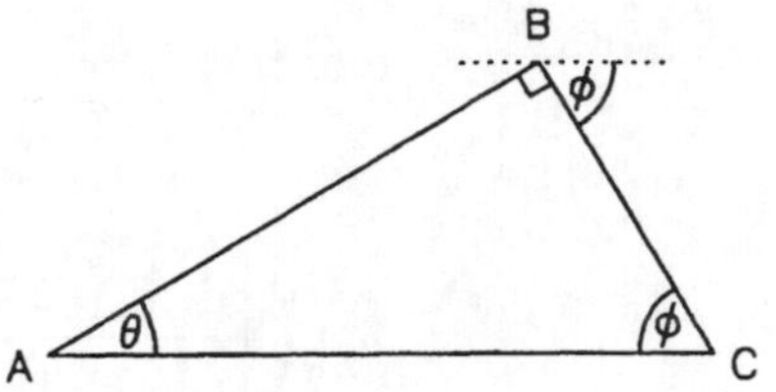

i.e. Gradient of line perpendicular to $y = mx + c$ is $-1/m$.

There is an alternative more algebraic proof of this result, but it is somewhat longer than the geometric argument above. We let the coordinates of the point of intersection be (x_0, y_0), and those of two arbitrary points on the respective lines, with slopes m and μ, be (x_1, y_1) and (x_2, y_2). Then, from the definition of a gradient, and Pythagoras' theorem, we have

$$m = \frac{y_1 - y_0}{x_1 - x_0} \quad \text{and} \quad \mu = \frac{y_2 - y_0}{x_2 - x_0}, \quad \text{and}$$

$$(x_1-x_0)^2 + (y_1-y_0)^2 + (x_2-x_0)^2 + (y_2-y_0)^2 = (x_1-x_2)^2 + (y_1-y_2)^2$$

With a suitable expansion, and cancellation, of the last equation, it is not very difficult to show that the three relationships lead to the result $\mu = -1/m$.

4.3 By exploiting the linearity property of a differential operator, and using a knowledge of the sum of an infinite GP, show that

$$\sum_{n=1}^{\infty} n x^{n-1} = \frac{\mathrm{d}}{\mathrm{d}x}\left(\frac{x}{1-x}\right)$$

where $|x|<1$. Hence, evaluate the summation.

$$\sum_{n=1}^{\infty} n x^{n-1} = \sum_{n=1}^{\infty} \frac{\mathrm{d}}{\mathrm{d}x}(x^n) = \frac{\mathrm{d}}{\mathrm{d}x}\left(\sum_{n=1}^{\infty} x^n\right) = \frac{\mathrm{d}}{\mathrm{d}x}\left(\frac{x}{1-x}\right) \quad \text{for } |x|<1$$

$$\therefore \quad \sum_{n=1}^{\infty} n x^{n-1} = \frac{1-x+x}{(1-x)^2} = \underline{\frac{1}{(1-x)^2}}$$

4.4 Differentiate $y=\cos^{-1}(x/a)$, where $|x/a|<1$; is the result valid for all values of y? What is the derivative of $\tan^{-1}(x/a)$?

$$\text{If} \quad y=\cos^{-1}(x/a), \quad \text{then } x = a\cos y \quad (\text{for } |x/a|<1)$$

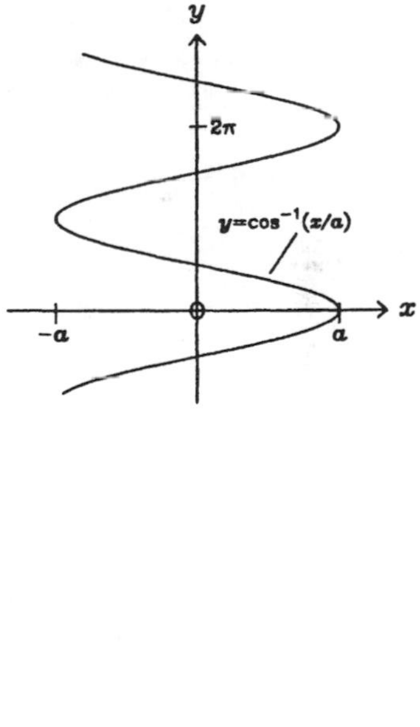

$$\therefore \quad \frac{\mathrm{d}x}{\mathrm{d}y} = -a\sin y = -a\sqrt{1-\cos^2 y} = -a\sqrt{1-x^2/a^2} = -\sqrt{a^2-x^2}$$

$$\therefore \quad \frac{\mathrm{d}y}{\mathrm{d}x} = \frac{1}{\mathrm{d}x/\mathrm{d}y} = \frac{-1}{\sqrt{a^2-x^2}}$$

$$\text{i.e.} \quad \underline{\frac{\mathrm{d}}{\mathrm{d}x}\left[\cos^{-1}(x/a)\right] = \frac{-1}{\sqrt{a^2-x^2}}}$$

This formula is correct for $0 \le y \le \pi$, and for multiple additions of 2π; in other words, for $2n\pi \le y \le (2n+1)\pi$ where n is an integer. If $\pi < y < 2\pi$, for example, then the derivative of $\cos^{-1}(x/a)$ is $+1/\sqrt{a^2-x^2}$.

$$y=\tan^{-1}(x/a) \quad \Longleftrightarrow \quad x = a\tan y$$

$$\therefore \quad \frac{\mathrm{d}x}{\mathrm{d}y} = a\sec^2 y = a\left(1+\tan^2 y\right) = a\left(1+x^2/a^2\right) = \left(a^2+x^2\right)/a$$

$$\therefore \quad \underline{\frac{\mathrm{d}}{\mathrm{d}x}\left[\tan^{-1}(x/a)\right] = \frac{a}{a^2+x^2}}$$

4.5 Differentiate the following functions, $y = f(x)$, with respect to x.

(a) $y = (2x+1)^3 \quad \Rightarrow \quad \frac{dy}{dx} = 3(2x+1)^2 \times 2$

$$\therefore \quad \frac{d}{dx}\left[(2x+1)^3\right] = 6(2x+1)^2$$

In this question, we have implicitly made the substitution $u = 2x+1$, so that $y = u^3$, and used the chain rule $dy/dx = dy/du \times du/dx$; the above result then follows from $dy/du = 3u^2$ and $du/dx = 2$.

(b) $y = \sqrt{3x-1} = (3x-1)^{1/2} \quad \Rightarrow \quad \frac{dy}{dx} = \frac{1}{2}(3x-1)^{-1/2} \times 3$

$$\therefore \quad \frac{d}{dx}\left(\sqrt{3x-1}\right) = \frac{3}{2\sqrt{3x-1}}$$

(c) $y = \cos 5x \quad \Rightarrow \quad \frac{dy}{dx} = -\sin 5x \times 5$

$$\therefore \quad \frac{d}{dx}(\cos 5x) = -5 \sin 5x$$

(d) $y = \sin(3x^2+7) \quad \Rightarrow \quad \frac{dy}{dx} = \cos(3x^2+7) \times (6x)$

$$\therefore \quad \frac{d}{dx}\left[\sin(3x^2+7)\right] = 6x\cos(3x^2+7)$$

(e) $y = \tan^4(2x+3) \quad \Rightarrow \quad \frac{dy}{dx} = 4\tan^3(2x+3) \times \sec^2(2x+3) \times 2$

$$\therefore \quad \frac{d}{dx}\left[\tan^4(2x+3)\right] = 8\tan^3(2x+3)\sec^2(2x+3)$$

Here we have implicitly used a slightly extended version of the chain rule $dy/dx = dy/du \times du/dv \times dv/dx$ where $v = 2x+3$, $u = \tan v$ and $y = u^4$; the result then follows from the substitution of the derivatives $dy/du = 4u^3$, $du/dv = \sec^2 v$ and $dv/dx = 2$.

(f) $y = x\,e^{-3x^2} \quad \Rightarrow \quad \frac{dy}{dx} = x\,e^{-3x^2} \times (-6x) + e^{-3x^2}$

$$\therefore \quad \frac{d}{dx}\left(x\,e^{-3x^2}\right) = (1-6x^2)\,e^{-3x^2}$$

This question involves the use of the product rule, $d/dx(uv) = u\,dv/dx + v\,du/dx$, where $u = x$ and $v = e^{-3x^2}$; the evaluation of dv/dx itself entails the chain rule, with $w = -3x^2$ and $v = e^w$.

(g) $y = x\ln(x^2+1) \;\Rightarrow\; \frac{dy}{dx} = x \times \frac{1}{(x^2+1)} \times 2x + \ln(x^2+1)$

$$\therefore\quad \frac{d}{dx}\left[x\ln(x^2+1)\right] = \frac{2x^2}{(x^2+1)} + \ln(x^2+1)$$

(h) $y = \frac{\sin x}{x} \;\Rightarrow\; \frac{dy}{dx} = \frac{x\cos x - \sin x}{x^2}$

$$\therefore\quad \frac{d}{dx}\left(\frac{\sin x}{x}\right) = \frac{x\cos x - \sin x}{x^2}$$

This is an example of the use of the quotient rule, $d/dx(u/v) = (v\,du/dx - u\,dv/dx)/v^2$, with $u = \sin x$ and $v = x$.

4.6 By first taking logarithms, differentiate $y = a^x$ where a is a constant.

$$y = a^x \;\Rightarrow\; \ln y = \ln(a^x) - x\ln a \qquad \text{— (1)}$$

$$\therefore\quad \frac{d}{dx}(1) \;\Rightarrow\; \frac{1}{y}\frac{dy}{dx} = \ln a$$

$$\therefore\quad \frac{dy}{dx} = y\ln a = a^x\ln a$$

$$\text{i.e.}\quad \frac{d}{dx}(a^x) = a^x\ln a$$

A simple verification of this formula is provided by the special case of $a = e$, whence $d/dx(e^x) = e^x$ is recovered because $\ln e = 1$.

4.7 Find dy/dx when: **(a)** $x = t(t^2+2)$ and $y = t^2$; **(b)** $x^2 = y\sin(xy)$.

(a)

$$\left.\begin{aligned} x = t^3 + 2t \;&\Rightarrow\; \frac{dx}{dt} = 3t^2 + 2 \\ y = t^2 \;&\Rightarrow\; \frac{dy}{dt} = 2t \end{aligned}\right\} \quad \therefore\quad \frac{dy}{dx} = \frac{dy/dt}{dx/dt} = \frac{2t}{(3t^2+2)}$$

(b) $x^2 = y\sin(xy) \;\Rightarrow\; \frac{d}{dx}(x^2) = \frac{d}{dx}[y\sin(xy)]$

$$\therefore\quad 2x = y\frac{d}{dx}[\sin(xy)] + \sin(xy)\frac{d}{dx}(y)$$

$$= y\cos(xy)\frac{d}{dx}(xy) + \sin(xy)\frac{dy}{dx}$$

$$= y\cos(xy)\left(x\frac{dy}{dx} + y\right) + \sin(xy)\frac{dy}{dx}$$

$$\therefore\quad 2x = y^2\cos(xy) + \frac{dy}{dx}[xy\cos(xy) + \sin(xy)]$$

$$\text{i.e.}\quad \underline{\frac{dy}{dx} = \frac{2x - y^2\cos(xy)}{xy\cos(xy) + \sin(xy)}}$$

4.8 Find and classify the stationary points of the following functions of x and r, where ε, σ and a_0 are constants.

(a) $f(x) = \frac{x^5}{5} - \frac{x^4}{6} - x^3 \qquad \therefore\; f'(x) = x^4 - \frac{2x^3}{3} - 3x^2$

$$\therefore\; f''(x) = 4x^3 - 2x^2 - 6x$$

For stationary points, $f'(x) = 0 \;\Rightarrow\; \frac{1}{3}x^2(3x^2 - 2x - 9) = 0$

$$\therefore\quad x = 0 \quad\text{or}\quad x = \frac{2 \pm \sqrt{4+108}}{6} = \frac{1 \pm 2\sqrt{7}}{3}$$

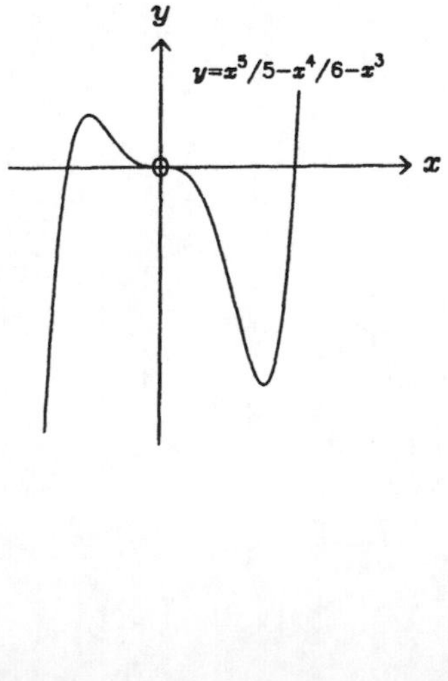

When $x = 0$, $f''(x) = 0 \qquad \therefore\; f''(x)$ test is inconclusive

As $x \to 0$, $f(x) \to -x^3 \qquad \therefore\; \underline{x = 0 \text{ is a point of inflexion}}$

When $x = \frac{1-2\sqrt{7}}{3}$, $f''(x) < 0 \qquad \therefore\; \underline{x = \frac{1-2\sqrt{7}}{3} \text{ is a maximum}}$

When $x = \frac{1+2\sqrt{7}}{3}$, $f''(x) > 0 \qquad \therefore\; \underline{x = \frac{1+2\sqrt{7}}{3} \text{ is a minimum}}$

(b) $\mathrm{f}(x) = \dfrac{x}{1+x^2}$

$$\therefore \ \mathrm{f}'(x) = \frac{1+x^2-2x^2}{(1+x^2)^2} = \frac{1-x^2}{(1+x^2)^2}$$

$$\therefore \ \mathrm{f}''(x) = \frac{-2x(1+x^2)^2-4x(1-x^2)(1+x^2)}{(1+x^2)^4} = \frac{-2x(3-x^2)}{(1+x^2)^3}$$

For stationary points, $\mathrm{f}'(x)=0 \ \Rightarrow \ 1-x^2=0$

$$\therefore \ x = \pm 1$$

$\mathrm{f}''(1) = -\frac{1}{2} < 0 \qquad \therefore$ $\underline{x=1 \text{ is a maximum}}$

$\mathrm{f}''(-1) = \frac{1}{2} > 0 \qquad \therefore$ $\underline{x=-1 \text{ is a minimum}}$

(c) $\mathrm{U}(r) = 4\varepsilon\left[\left(\frac{\sigma}{r}\right)^{12} - \left(\frac{\sigma}{r}\right)^{6}\right]$ (for $r \geq 0$)

$$\therefore \ \mathrm{U}'(r) = 4\varepsilon\left[-12\left(\tfrac{\sigma}{r}\right)^{11}\tfrac{\sigma}{r^2} + 6\left(\tfrac{\sigma}{r}\right)^{5}\tfrac{\sigma}{r^2}\right] = \tfrac{24\varepsilon}{\sigma}\left[\left(\tfrac{\sigma}{r}\right)^{7} - 2\left(\tfrac{\sigma}{r}\right)^{13}\right]$$

$$\therefore \ \mathrm{U}''(r) = \tfrac{24\varepsilon}{\sigma}\left[-7\left(\tfrac{\sigma}{r}\right)^{6}\tfrac{\sigma}{r^2} + 26\left(\tfrac{\sigma}{r}\right)^{12}\tfrac{\sigma}{r^2}\right] = \tfrac{24\varepsilon}{\sigma^2}\left[26\left(\tfrac{\sigma}{r}\right)^{14} - 7\left(\tfrac{\sigma}{r}\right)^{8}\right]$$

For stationary points, $\mathrm{U}'(r)=0 \ \Rightarrow \ \left(\frac{\sigma}{r}\right)^{7}\left[1-2\left(\frac{\sigma}{r}\right)^{6}\right]=0$

$$\therefore \ r \to \infty \ \text{ or } \ r = 2^{1/6}\sigma$$

As $r \to \infty$, $\mathrm{U}(r)$ decays to zero $\therefore$ Not a proper stationary point.

$\mathrm{U}''(2^{1/6}\sigma) > 0 \qquad \therefore$ $\underline{r = 2^{1/6}\sigma \text{ is a minimum}}$

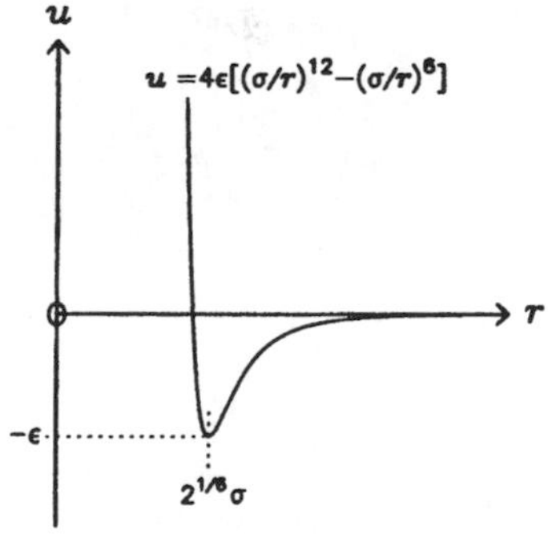

The function $\mathrm{U}(r)$ is known as the *Lennard-Jones 6-12 potential*, and is often used to approximate the potential energy between molecules. While there are no interactions between the constituents of an *ideal gas*, there are in real ones: a short-range repulsion, which stops two molecules from occupying the same space; and a longer-range attraction (from a van der Waals type dipole-induced-dipole interaction) which weakens as the molecules move further apart. The optimal separation r represents a balance between these opposing forces, and is given by the location of the minimum in $\mathrm{U}(r)$.

(d) $\quad \mathrm{p}(r) = \frac{r^2}{8a_0^3}\left(2 - \frac{r}{a_0}\right)^2 \mathrm{e}^{-r/a_0} \qquad$ (for $r \geq 0$)

$$\therefore \ \mathrm{p}'(r) = \frac{1}{8a_0^3}\left[-\frac{r^2}{a_0}\left(2-\frac{r}{a_0}\right)^2 + 2r\left(2-\frac{r}{a_0}\right)^2 - \frac{2r^2}{a_0}\left(2-\frac{r}{a_0}\right)\right]\mathrm{e}^{-r/a_0}$$

$$= \frac{r}{8a_0^3}\left(2-\frac{r}{a_0}\right)\left[\left(\frac{r}{a_0}\right)^2 - 6\frac{r}{a_0} + 4\right]\mathrm{e}^{-r/a_0}$$

$$\therefore \ \mathrm{p}''(r) = \frac{1}{8a_0^3}\left\{\left(2-\frac{r}{a_0}\right)\left[\left(\frac{r}{a_0}\right)^2 - 6\frac{r}{a_0} + 4\right]\right.$$

$$-\frac{r}{a_0}\left[\left(\frac{r}{a_0}\right)^2 - 6\frac{r}{a_0} + 4\right]$$

$$+\frac{r}{a_0}\left(2-\frac{r}{a_0}\right)\left[2\frac{r}{a_0} - 6\right]$$

$$\left.-\frac{r}{a_0}\left(2-\frac{r}{a_0}\right)\left[\left(\frac{r}{a_0}\right)^2 - 6\frac{r}{a_0} + 4\right]\right\}\mathrm{e}^{-r/a_0}$$

$$= \frac{1}{8a_0^3}\left[\left(\frac{r}{a_0}\right)^4 - 12\left(\frac{r}{a_0}\right)^3 + 40\left(\frac{r}{a_0}\right)^2 - 40\left(\frac{r}{a_0}\right) + 8\right]\mathrm{e}^{-r/a_0}$$

For stationary points, $\mathrm{p}'(r) = 0$

$$\therefore \quad r = 0, \quad r = 2a_0, \quad \frac{r}{a_0} = \frac{6 \pm \sqrt{36-16}}{2} = 3 \pm \sqrt{5}, \quad \text{or} \quad r \to \infty$$

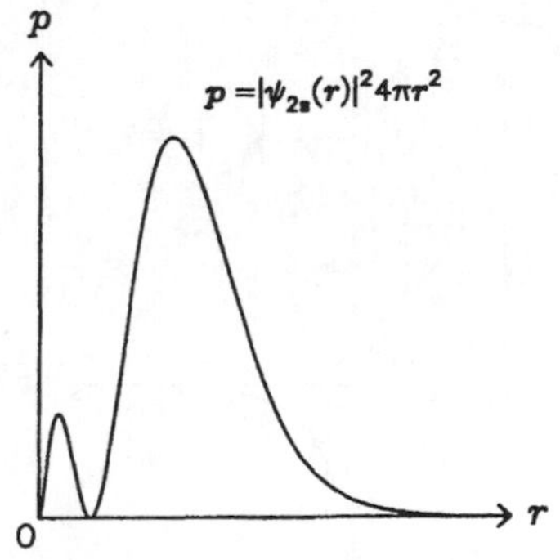

When $r = 0$ or $r \to \infty$, they are not proper stationary points.

$\mathrm{p}''(2a_0) > 0 \qquad \therefore$ <u>$r = 2a_0$ is a minimum</u>

$\mathrm{p}''[(3-\sqrt{5})a_0] < 0 \qquad \therefore$ <u>$r = (3-\sqrt{5})a_0$ is a maximum</u>

$\mathrm{p}''[(3+\sqrt{5})a_0] < 0 \qquad \therefore$ <u>$r = (3+\sqrt{5})a_0$ is a maximum</u>

The function $\mathrm{p}(r)$ is the radial probability density function for an electron in the $2s$ state of a hydrogen atom; that is, the probability of finding the electron in a small interval between r and $r + \delta r$ from the nucleus is given by $\mathrm{p}(r)\,\delta r$ (where $\mathrm{p}(r) = |\psi_{2s}|^2 4\pi r^2$). In addition to the main maximum at $r = (3+\sqrt{5})\,a_0$, where the electron is most probably to be found, there is also a smaller maximum closer to the nucleus at $r = (3-\sqrt{5})\,a_0$; this subsidiary maximum is indicative of the *penetration effect*, and gives an important insight into the behaviour of electrons in atomic shells.

5 Integration

5.1 Integrate the following functions with respect to x:
(a) $x+\sqrt{x}-1/x$; **(b)** $\sqrt{x}\,(x-1/x)$; **(c)** 2^x; **(d)** e^{2x}; **(e)** $1/(2x-1)$;
(f) $\sin 2x+\cos 3x$; **(g)** $\tan x$; **(h)** $\sin^2 x$; **(i)** $x/(1+x^2)$; **(j)** $1/(1+x^2)$.

(a)
$$\int \left(x+\sqrt{x}-\tfrac{1}{x}\right) dx = \int \left(x+x^{1/2}-x^{-1}\right) dx$$
$$= \frac{x^2}{2}+\frac{x^{3/2}}{3/2}-\ln x+C$$
$$= \underline{\frac{x^2}{2}+\frac{2x\sqrt{x}}{3}-\ln x+C}$$

(b)
$$\int \sqrt{x}\left(x-\tfrac{1}{x}\right) dx = \int \left(x^{3/2}-x^{-1/2}\right) dx$$
$$= \frac{x^{5/2}}{5/2}-\frac{x^{1/2}}{1/2}+C$$
$$= \underline{2\sqrt{x}\left(\tfrac{x^2}{5}-1\right)+C}$$

(c)
$$\int 2^x\, dx = \underline{\frac{2^x}{\ln 2}+C}$$

(d)
$$\int e^{2x}\, dx = \underline{\frac{e^{2x}}{2}+C}$$

(e)
$$\int \frac{dx}{2x-1} = \underline{\tfrac{1}{2}\ln(2x-1)+C}$$

If the integral above is not obvious, then the implicit manipulation can be made explicit with the substitution $u=2x-1$ (giving $du=2\,dx$):

$$\int \frac{dx}{2x-1} = \int \frac{du/2}{u} = \frac{1}{2}\int \frac{du}{u} = \frac{1}{2}\ln u+C$$

(f) $$\int (\sin 2x + \cos 3x)\,dx = -\tfrac{1}{2}\cos 2x + \tfrac{1}{3}\sin 3x + C$$

(g) $$\int \tan x\,dx = \int \frac{\sin x}{\cos x}\,dx$$

$$= -\ln(\cos x) + C = \ln(\sec x) + C$$

Having written $\tan x$ as $\sin x/\cos x$, the integral becomes easy because the numerator is equal to the derivative of the denominator (give or take a minus sign). If the significance of the latter is not apparent in terms of the chain rule of differentiation, then the final result can be obtained with the substitution $u = \cos x$ (whence $du = -\sin x\,dx$):

$$\int \frac{\sin x}{\cos x}\,dx = -\int \frac{du}{u} = -\ln u + C$$

(h) $$\int \sin^2 x\,dx = \tfrac{1}{2}\int (1 - \cos 2x)\,dx$$

$$= \tfrac{1}{2}\left(x - \frac{\sin 2x}{2}\right) + C$$

$$= \frac{x}{2} - \frac{\sin 2x}{4} + C$$

(i) $$\int \frac{x}{1+x^2}\,dx = \tfrac{1}{2}\ln(1+x^2) + C$$

This is again an easy integral because the numerator is equal to the derivative of the denominator, to within a constant multiplicative factor; the formal substitution, if required, would be $u = 1 + x^2$ (and $du = 2x\,dx$):

(j) $$\int \frac{dx}{1+x^2} = \tan^{-1}x + C$$

Although we have simply written the answer to this one straight down, because it is a 'standard integral', it can be ascertained systematically by substituting $x = \tan\theta$. Then, using the derivative $dx = \sec^2\theta\,d\theta = (1 + \tan^2\theta)\,d\theta = (1 + x^2)\,d\theta$ we are led to the result above:

$$\int \frac{dx}{1+x^2} = \int d\theta = \theta + C = \tan^{-1}x + C$$

5.2 Evaluate the following definite integrals:

$$\text{(a)}\ \int_0^{\pi/2} \sin^4 x \cos x \, dx\,, \qquad \text{(b)}\ \int_0^{\pi/2} \sin^4 x \, dx\,, \qquad \text{(c)}\ \int_0^4 \frac{x+3}{\sqrt{2x+1}} \, dx$$

(a)

$$\int_0^{\pi/2} \sin^4 x \cos x \, dx = \left[\tfrac{1}{5}\sin^5 x\right]_0^{\pi/2}$$

$$= \tfrac{1}{5}\left[\sin^5\left(\tfrac{\pi}{2}\right) - \sin^5(0)\right]$$

$$= \underline{\tfrac{1}{5}}$$

The integral of $\sin^4 x \cos x$ is easy to write down because of the presence of a $\cos x$ with the function of $\sin x$; the manipulation mentally carried out above can be made explicit with substitution $u = \sin x$ and $du = \cos x\, dx$, leading to $\int \sin^4 x \cos x\, dx = \int u^4\, du$. If the derivative of $\sin x$ had not been present in the integrand, then the sum would have entailed a lot more effort; in particular, the (repeated) use of the double-angle formula $\cos 2x = 1 - 2\sin^2 x$.

(b) From exercise 3.6, we have $8\sin^4\theta = \cos 4\theta - 4\cos 2\theta + 3$.

$$\therefore \quad \int_0^{\pi/2} \sin^4 x \, dx = \tfrac{1}{8}\int_0^{\pi/2} (\cos 4x - 4\cos 2x + 3)\, dx$$

$$= \tfrac{1}{8}\left[\tfrac{1}{4}\sin 4x - 2\sin 2x + 3x\right]_0^{\pi/2}$$

$$= \tfrac{1}{8}\left[\tfrac{1}{4}(\sin 2\pi - \sin 0) - 2(\sin\pi - \sin 0) + 3\left(\tfrac{\pi}{2} - 0\right)\right]$$

$$= \underline{\frac{3\pi}{16}}$$

(c) Let $\mathrm{I} = \displaystyle\int_0^4 \frac{x+3}{\sqrt{2x+1}}\, dx$

Put $u^2 = 2x+1 \qquad \therefore\ 2u\, du = 2\, dx$

$$\therefore \quad I = \int_{u=1}^{u=3} \frac{[(u^2-1)/2+3]}{u}\, u\, du = \int_1^3 \left(\tfrac{u^2}{2} - \tfrac{1}{2} + 3\right) du$$

$$= \left[\frac{u^3}{6} + \frac{5u}{2}\right]_1^3$$

$$= \tfrac{27}{6} + \tfrac{15}{2} - \tfrac{1}{6} - \tfrac{5}{2}$$

$$= \tfrac{13}{3} + 5 = \underline{9\tfrac{1}{3}}$$

An alternative way of proceeding with this calculation is to use 'integration by parts', where $(2x+1)^{-1/2}$ is integrated and $x+3$ is differentiated:

$$\int_0^4 \frac{x+3}{\sqrt{2x+1}}\, dx = \left[(x+3)\sqrt{2x+1}\right]_0^4 - \int_0^4 \sqrt{2x+1}\, dx$$

$$= 21 - 3 - \left[\tfrac{1}{3}(2x+1)^{3/2}\right]_0^4$$

$$= 18 - \tfrac{1}{3}[9^{3/2} - 1]$$

$$= 18 - \tfrac{26}{3} = \underline{9\tfrac{1}{3}}$$

5.3 Integrate the three partial fraction expressions in exercise 1.9, with respect to x.

(a) $$\int \frac{dx}{x^2-5x+6} = \int \frac{dx}{x-3} - \int \frac{dx}{x-2}$$

$$= \ln(x-3) - \ln(x-2) + C$$

$$= \underline{\ln\left(\frac{x-3}{x-2}\right) + C}$$

(b) $$\int \frac{x^2-5x+1}{(x-1)^2(2x-3)}\, dx = 3\int \frac{dx}{(x-1)^2} + 9\int \frac{dx}{x-1} - 17\int \frac{dx}{2x-3}$$

$$= \underline{C - \frac{3}{x-1} + 9\ln(x-1) - \tfrac{17}{2}\ln(2x-3)}$$

(c) $\displaystyle\int \frac{11x+1}{(x-1)(x^2-3x-2)}\,dx = \int \frac{3x+5}{x^2-3x-2}\,dx - 3\int \frac{dx}{x-1}$

$$= \frac{3}{2}\int \frac{2x-3}{x^2-3x-2}\,dx + \frac{19}{2}\int \frac{dx}{x^2-3x-2} - 3\int \frac{dx}{x-1}$$

$$= \frac{3}{2}\ln(x^2-3x-2) + \frac{19}{2}\int \frac{dx}{(x-3/2)^2-(\sqrt{17}/2)^2} - 3\ln(x-1) + C$$

$$= \underline{3\ln\left(\frac{\sqrt{x^2-3x-2}}{x-1}\right) - \frac{19}{\sqrt{17}}\tanh^{-1}\left(\frac{2x-3}{\sqrt{17}}\right) + K}$$

Like many cases in real life, the evaluation of this integral has involved several manipulations. Starting with a partial fraction decomposition, transcribed from exercise 9.1, and followed by the rewriting of the numerator $3x+5$ as $3(2x-3)/2+19/2$, so that the first part becomes a recognisable integral, to the completion of the squares in the denominator x^3-3x-2; this last step allows us to exploit the 'standard result' that $\mathrm{d}/\mathrm{d}\theta\left[\tanh^{-1}(\theta/a)\right] = a/(a^2-\theta^2)$. We could also have written $\tanh^{-1}(\theta/a)$ as $\ln\left[(a+\theta)/(a-\theta)\right]/2$, or derived this form of the integral by expressing the denominator θ^2-a^2 as the product $(\theta-a)(\theta+a)$, where $\theta=(x-3/2)$ and $a=\sqrt{17}/2$, and using partial fractions.

5.4 By integrating by parts, show that

(a) $\displaystyle\int x\sin x\,dx = C - x\cos x + \sin x\,,$ and

(b) $\displaystyle\int \sin x\, e^{-x}\,dx = C - \tfrac{1}{2}(\sin x+\cos x)\,e^{-x}$

(a) $\displaystyle\int x\sin x\,dx = -x\cos x + \int \cos x\,dx$

$$= \underline{-x\cos x + \sin x + C}$$

(b) Let $\displaystyle I = \int \sin x\, e^{-x}\,dx$

$\therefore\quad \displaystyle I = -e^{-x}\sin x + \int e^{-x}\cos x\,dx$

$$= -e^{-x}\sin x - e^{-x}\cos x - \int e^{-x}\sin x\,dx$$

$$\therefore \quad I = -(\sin x + \cos x)e^{-x} - I$$

$$\text{i.e.} \quad \underline{\int \sin x \, e^{-x} \, dx = C - \tfrac{1}{2}(\sin x + \cos x)e^{-x}}$$

5.5 If

$$I_n = \int_0^{\pi/2} \sin^n x \, dx ,$$

for $n \geq 1$, show that $n I_n = (n-1) I_{n-2}$. Hence, by relating odd orders of I_n to I_1 and even ones to I_0, evaluate I_5 and I_8.

$$I_n = \int_0^{\pi/2} \sin^{n-1} x \sin x \, dx \qquad (\text{for } n \geq 1)$$

$$= \Big[-\sin^{n-1} x \cos x\Big]_0^{\pi/2} + (n-1)\int_0^{\pi/2} \sin^{n-2} x \cos^2 x \, dx$$

$$= 0 + (n-1)\int_0^{\pi/2} \sin^{n-2} x \, (1 - \sin^2 x) \, dx$$

$$= (n-1)\left\{ \int_0^{\pi/2} \sin^{n-2} x \, dx - \int_0^{\pi/2} \sin^n x \, dx \right\}$$

$$= (n-1)(I_{n-2} - I_n)$$

$$\therefore \quad \frac{I_n}{(n-1)} + I_n = I_{n-2}$$

$$\text{i.e.} \quad \underline{n I_n = (n-1) I_{n-2}} \quad \text{for } n \geq 1$$

$$I_n = \left(\frac{n-1}{n}\right) I_{n-2}$$

$$\therefore \quad I_5 = \tfrac{4}{5} I_3 = \tfrac{4}{5} \times \tfrac{2}{3} I_1$$

$$\text{But} \quad I_1 = \int_0^{\pi/2} \sin x \, dx = \Big[-\cos x\Big]_0^{\pi/2} = 1 \qquad \therefore \quad \underline{I_5 = \tfrac{8}{15}}$$

$$I_8 = \tfrac{7}{8}\, I_6 = \tfrac{7}{8} \times \tfrac{5}{6}\, I_4 = \tfrac{7}{8} \times \tfrac{5}{6} \times \tfrac{3}{4} \times \tfrac{1}{2}\, I_0$$

But $I_0 = \int_0^{\pi/2} dx = \left[x\right]_0^{\pi/2} = \frac{\pi}{2}$ $\qquad \therefore\ \underline{I_8 = \frac{35\pi}{256}}$

5.6 Prove algebraically that $\int_{-a}^{a} f(x)\,dx = 0$ if $f(-x) = -f(x)$, and derive the corresponding expression for the symmetric function.

$$\int_{-a}^{a} f(x)\,dx = \int_{-a}^{0} f(x)\,dx + \int_{0}^{a} f(x)\,dx$$

$$= -\int_{a}^{0} f(-u)\,du + \int_{0}^{a} f(x)\,dx \qquad \text{(where } x = -u\text{)}$$

$$= \int_{a}^{0} f(u)\,du + \int_{0}^{a} f(x)\,dx \qquad [f(-u) = -f(u)]$$

$$= -\int_{0}^{a} f(u)\,du + \int_{0}^{a} f(x)\,dx = \underline{0}$$

In this example, we began by splitting up the integral from $-a$ to $+a$ into two parts (for positive x and negative x); then we made the substitution $u = -x$ (so that $du = -dx$), used the fact that the integrand was antisymmetric, and interchanged the order of the limits. This eventually led to the difference of two identical integrals, even though one was written in terms of x and the other in u. The reason that the x and u discrepancy doesn't matter is because they are 'dummy variables'; that is to say, the result of the definite integral is purely a number (dependent on a) and not on a function of x or u.

If the integrand is symmetric, so that $f(-x) = f(x)$, then a retracing of the steps above leads to

$$\int_{-a}^{a} f(x)\,dx = \int_{0}^{a} f(u)\,du + \int_{0}^{a} f(x)\,dx = \underline{2\int_{0}^{a} f(x)\,dx}$$

Incidentally, any function can be rewritten as the sum of a symmetric (or even) contribution and an antisymmetric (or odd) one:

$$f(x) = \underbrace{\tfrac{1}{2}\left[f(x) + f(-x)\right]}_{f_{\text{even}}(x)} + \underbrace{\tfrac{1}{2}\left[f(x) - f(-x)\right]}_{f_{\text{odd}}(x)}$$

6 Taylor series

6.1 Derive the Taylor series for $\sin(x+\pi/6)$ for small x.

Let $\mathrm{f}(x) = \sin x \qquad \therefore \quad \mathrm{f}(\pi/6) = 1/2$

$\therefore \quad \mathrm{f}'(x) = \cos x \qquad \mathrm{f}'(\pi/6) = \sqrt{3}/2$

$\mathrm{f}''(x) = -\sin x \qquad \mathrm{f}''(\pi/6) = -1/2$

$\mathrm{f}'''(x) = -\cos x \qquad \mathrm{f}'''(\pi/6) = -\sqrt{3}/2$

$\mathrm{f}''''(x) = \sin x \qquad \mathrm{f}''''(\pi/6) = 1/2$

But $\mathrm{f}(x+a) = \mathrm{f}(a) + x\,\mathrm{f}'(a) + \frac{x^2}{2!}\mathrm{f}''(a) + \frac{x^3}{3!}\mathrm{f}'''(a) + \cdots$

$$\therefore \quad \underline{\sin(x+\tfrac{\pi}{6}) = \tfrac{1}{2} + \tfrac{\sqrt{3}}{2}x - \tfrac{1}{4}x^2 - \tfrac{\sqrt{3}}{12}x^3 + \tfrac{1}{48}x^4 + \cdots}$$

6.2 Derive the binomial expansion for $(1+x)^n$. By expressing $\sqrt{C}$ as $a(1+b)^{1/2}$, where a and b are suitable constants, find $\sqrt{8}$ and $\sqrt{17}$ to four decimal places.

Let $\mathrm{f}(x) = x^n \qquad \therefore \quad \mathrm{f}(1) = 1$

$\therefore \quad \mathrm{f}'(x) = nx^{n-1} \qquad \mathrm{f}'(1) = n$

$\mathrm{f}''(x) = n(n-1)x^{n-2} \qquad \mathrm{f}''(1) = n(n-1)$

$\mathrm{f}'''(x) = n(n-1)(n-2)x^{n-3} \qquad \mathrm{f}'''(1) = n(n-1)(n-2)$

But $\mathrm{f}(a+x) = \mathrm{f}(a) + x\,\mathrm{f}'(a) + \frac{x^2}{2!}\mathrm{f}''(a) + \frac{x^3}{3!}\mathrm{f}'''(a) + \cdots$

$$\therefore \quad \underline{(1+x)^n = 1 + nx + \frac{n(n-1)}{2}x^2 + \frac{n(n-1)(n-2)}{6}x^3 + \cdots} \quad \text{for } |x|<1$$

$$n = \tfrac{1}{2} \quad \Rightarrow \quad \sqrt{1+x} = 1 + \tfrac{1}{2}x - \tfrac{1}{8}x^2 + \tfrac{1}{16}x^3 - \tfrac{5}{128}x^4 + \cdots \quad \text{for } |x|<1$$

$$\sqrt{8} = \sqrt{9-1} = 3\sqrt{1-1/9}$$

$$\therefore\ \sqrt{8} = 3\left[1+\tfrac{1}{2}\left(-\tfrac{1}{9}\right)-\tfrac{1}{8}\left(-\tfrac{1}{9}\right)^2+\tfrac{1}{16}\left(-\tfrac{1}{9}\right)^3-\tfrac{5}{128}\left(-\tfrac{1}{9}\right)^4+\cdots\right]$$

$$= 3\left[1-\tfrac{1}{18}-\tfrac{1}{648}-\tfrac{1}{11664}-\tfrac{5}{839808}-\cdots\right]$$

$$= \underline{2.8284} \quad \text{to 4 decimal places}$$

$$\sqrt{17} = \sqrt{16+1} = 4\sqrt{1+1/16}$$

$$\therefore\ \sqrt{17} = 4\left[1+\tfrac{1}{2}\left(\tfrac{1}{16}\right)-\tfrac{1}{8}\left(\tfrac{1}{16}\right)^2+\tfrac{1}{16}\left(\tfrac{1}{16}\right)^3-\cdots\right]$$

$$= 4\left[1+\tfrac{1}{32}-\tfrac{1}{2048}-\tfrac{1}{65536}+\cdots\right]$$

$$= \underline{4.1231} \quad \text{to 4 decimal places}$$

The binomial expansion converges increasingly quickly as $|x| \to 0$. Thus, for example, in the calculation of $\sqrt{1.0000001} = (1+10^{-7})^{1/2}$, the inclusion of the x^2 term (which is -1.25×10^{-15}) yields greater accuracy than that provided by most electronic calculators! More seriously, though, a sum like $\sqrt{a^2+b^2}$ where $a \gg b$ is best done by using the binomial expansion

$$\sqrt{a^2+b^2} = a\left[1+\left(\frac{b}{a}\right)^2\right]^{1/2} = a\left[1+\tfrac{1}{2}\left(\frac{b}{a}\right)^2-\tfrac{1}{8}\left(\frac{b}{a}\right)^4+\tfrac{1}{16}\left(\frac{b}{a}\right)^6-\cdots\right]$$

so as to avoid 'rounding errors', even in a high precision computer, when $|b/a| \ll 1$.

> **6.3** Derive the Taylor series for $\ln(1+x)$ and $\ln(1-x)$. Hence, write down the power series for $\ln\left[(1+x)/(1-x)\right]$.

Let $f(x) = \ln x$	$\therefore\ f(1) = 0$
$\therefore\ f'(x) = x^{-1}$	$f'(1) = 1$
$f''(x) = -x^{-2}$	$f''(1) = -1$
$f'''(x) = 2x^{-3}$	$f'''(1) = 2$
$f''''(x) = -6x^{-4}$	$f''''(1) = -6$
$f'''''(x) = 24x^{-5}$	$f'''''(1) = 24$

But $\mathrm{f}(a+x)=\mathrm{f}(a)+x\,\mathrm{f}'(a)+\frac{x^2}{2!}\mathrm{f}''(a)+\frac{x^3}{3!}\mathrm{f}'''(a)+\cdots$

$$\therefore\quad \ln(1+x)=x-\frac{x^2}{2}+\frac{x^3}{3}-\frac{x^4}{4}+\frac{x^5}{5}-\cdots \quad \text{for } |x|<1$$

If we substitute $x=-u$ (say) in the above, then it follows that

$$\ln(1-x)=-x-\frac{x^2}{2}-\frac{x^3}{3}-\frac{x^4}{4}-\frac{x^5}{5}-\cdots \quad \text{for } |x|<1$$

$$\ln\left(\frac{1+x}{1-x}\right)=\ln(1+x)-\ln(1-x)$$

$$=2\left(x+\frac{x^3}{3}+\frac{x^5}{5}+\cdots\right) \quad \text{for } |x|<1$$

6.4 Derive the Maclaurin series for $\cos^{-1}x$.

The Maclaurin series is simply a special case of a Taylor series where the expansion is about the origin; that is to say, $a=0$. We could ascertain it in the usual systematic way by setting up a two-column list with $\mathrm{f}(x)=\cos^{-1}x$ and its derivatives on the left, and their values at $x=0$ on the right. An algebraically easier way, however, is to differentiate $\cos^{-1}x$ just once, use a binomial expansion and integrate term-by-term.

Let $\mathrm{f}(x)=\cos^{-1}x \qquad \therefore\quad \mathrm{f}'(x)=\frac{-1}{\sqrt{1-x^2}}=-\left(1-x^2\right)^{-1/2}$

But $\left(1-x^2\right)^{-1/2}=1+\frac{1}{2}x^2+\frac{3}{8}x^4+\frac{5}{16}x^6+\frac{35}{128}x^8+\cdots$

$$\therefore\quad \mathrm{f}(x)=\int \mathrm{f}'(x)\,\mathrm{d}x$$

$$=C-x-\frac{1}{6}x^3-\frac{3}{40}x^5-\frac{5}{112}x^7-\frac{35}{1152}x^9-\cdots$$

But $\cos^{-1}(0)=\frac{\pi}{2} \;\Rightarrow\; C=\frac{\pi}{2}$

$$\therefore\quad \cos^{-1}x=\frac{\pi}{2}-x-\frac{x^3}{6}-\frac{3x^5}{40}-\frac{5x^7}{112}-\cdots$$

We should remember that, like $\mathrm{d}/\mathrm{d}x(\cos^{-1}x)=-1/\sqrt{1-x^2}$, the Maclaurin series is only valid for $|x|<1$ and $0\le\cos^{-1}x\le\pi$.

6.5 Determine the following limits:

(a) $\lim_{x\to 0} \frac{\sin ax}{x}$, **(b)** $\lim_{x\to 0} \frac{\cos x - 1}{x}$, **(c)** $\lim_{x\to 0} \frac{2\cos x + x\sin x - 2}{x^4}$

(a)
$$\lim_{x\to 0} \frac{\sin ax}{x} = \lim_{x\to 0} \frac{ax - a^3x^3/6 + \cdots}{x}$$
$$= \lim_{x\to 0} a - \frac{a^3}{6}x^2 + \cdots$$
$$= \underline{a}$$

Alternatively, we could use L'Hospital's rule:

$$\lim_{x\to 0} \frac{\sin ax}{x} = \lim_{x\to 0} \frac{a\cos ax}{1}$$
$$= \underline{a}$$

(b)
$$\lim_{x\to 0} \frac{\cos x - 1}{x} = \lim_{x\to 0} \frac{(1 - x^2/2 + x^4/24 - \cdots) - 1}{x}$$
$$= \lim_{x\to 0} -\frac{x}{2} + \frac{x^3}{24} + \cdots$$
$$= \underline{0}$$

or $$\lim_{x\to 0} \frac{\cos x - 1}{x} = \lim_{x\to 0} \frac{-\sin x}{1} \qquad \text{(L'Hospital)}$$
$$= \underline{0}$$

(c)
$$\lim_{x\to 0} \frac{2\cos x + x\sin x - 2}{x^4}$$
$$= \lim_{x\to 0} \frac{2(1 - x^2/2 + x^4/24 - \cdots) + x(x - x^3/6 + \cdots) - 2}{x^4}$$
$$= \lim_{x\to 0} -\frac{1}{12} + O(x^2)$$
$$= \underline{-\frac{1}{12}}$$

The notation $O(x^2)$ means 'of order x^2'; that is to say, the omitted terms contain a factor of x^2 and other contributions with powers of x greater than two.

$$\text{or} \quad \lim_{x\to 0} \frac{2\cos x + x\sin x - 2}{x^4} = \lim_{x\to 0} \frac{-\sin x + x\cos x}{4x^3} \qquad \text{(L'Hospital)}$$

$$= \lim_{x\to 0} \frac{-x\sin x}{12x^2} \qquad \text{(L'Hospital)}$$

$$= \lim_{x\to 0} \frac{-\sin x}{12x}$$

$$= \lim_{x\to 0} \frac{-\cos x}{12} \qquad \text{(L'Hospital)}$$

$$= \underline{-\tfrac{1}{12}}$$

6.6 Given that $x^3+3x^2+6x-3=0$ has only one (real) root, find its value to five significant figures by using the Newton-Raphson method.

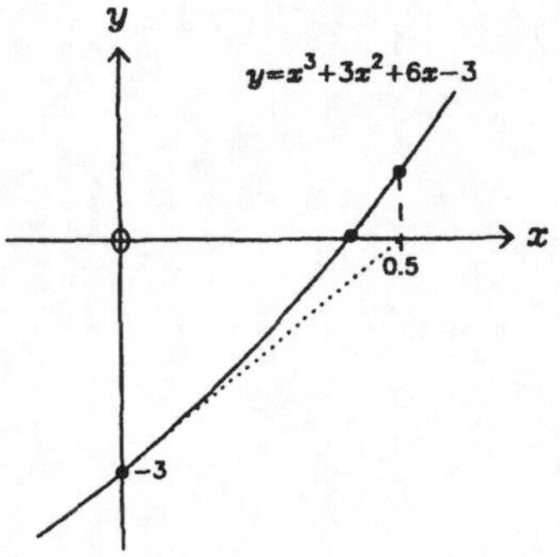

Let $f(x) = x^3+3x^2+6x-3 \qquad \therefore \quad f'(x) = 3x^2+6x+6$

If $f(x_n) \approx 0$, then a better estimate of $f(x) = 0$, x_{n+1}, is given by

$$x_{n+1} = x_n - \frac{f(x_n)}{f'(x_n)} = x_n - \frac{x_n^3 + 3x_n^2 + 6x_n - 3}{3x_n^2 + 6x_n + 6}$$

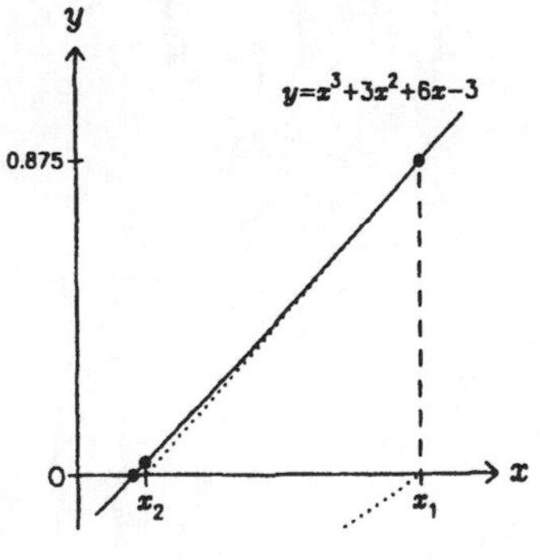

Let $x_0 = 0$	$\therefore \quad f(x_0) = -3.000$
$\therefore \quad x_1 = 0.5$	$f(x_1) = 0.875$
$x_2 = 0.4102564$	$f(x_2) = 0.03552$
$x_3 = 0.4062950$	$f(x_3) = 0.00006633$
$x_4 = 0.4062876$	$f(x_4) = 0$

i.e. $x^3+3x^2+6x-3=0$ when $\underline{x = 0.40629}$ (5 sig. fig.)

7 Complex numbers

7.1 If $u = 2+3\mathrm{i}$ and $v = 1-\mathrm{i}$, find the real and imaginary parts of: **(a)** $u+v$, **(b)** $u-v$, **(c)** uv, **(d)** u/v, and **(e)** v/u.

(a) $u+v = 2+1+\mathrm{i}(3-1) = \underline{3+2\mathrm{i}}$

(b) $u-v = 2-1+\mathrm{i}(3+1) = \underline{1+4\mathrm{i}}$

(c) $uv = (2+3\mathrm{i})(1-\mathrm{i}) = 2-2\mathrm{i}+3\mathrm{i}-3\mathrm{i}^2 = \underline{5+\mathrm{i}}$ $\qquad \mathrm{i}^2 = -1$

(d) $\frac{u}{v} = \frac{2+3\mathrm{i}}{1-\mathrm{i}} = \left(\frac{2+3\mathrm{i}}{1-\mathrm{i}}\right)\times\left(\frac{1+\mathrm{i}}{1+\mathrm{i}}\right) = \frac{2+2\mathrm{i}+3\mathrm{i}+3\mathrm{i}^2}{1+\mathrm{i}-\mathrm{i}-\mathrm{i}^2} = \underline{\frac{-1+5\mathrm{i}}{2}}$

(e) $\frac{v}{u} = \left(\frac{1-\mathrm{i}}{2+3\mathrm{i}}\right)\times\left(\frac{2-3\mathrm{i}}{2-3\mathrm{i}}\right) = \frac{2-3\mathrm{i}-2\mathrm{i}+3\mathrm{i}^2}{4+9} = \underline{\frac{-1-5\mathrm{i}}{13}}$

or $= \frac{1}{u/v} = \left(\frac{2}{-1+5\mathrm{i}}\right)\times\left(\frac{-1-5\mathrm{i}}{-1-5\mathrm{i}}\right) = \frac{2-10\mathrm{i}}{1+25} = \underline{\frac{-1-5\mathrm{i}}{13}}$

7.2 For the previous example, find: **(a)** $|u|$, **(b)** $|v|$, **(c)** $|uv|$, **(d)** $|u/v|$, and **(e)** $|v/u|$.

(a) $|u|^2 = uu^* = (2+3\mathrm{i})(2-3\mathrm{i}) = 4+9 \quad \therefore \quad \underline{|u| = \sqrt{13}}$

(b) $|v|^2 = vv^* = (1-\mathrm{i})(1+\mathrm{i}) = 1+1 \quad \therefore \quad \underline{|v| = \sqrt{2}}$

(c) $|uv|^2 = (uv)(uv)^* = (5+\mathrm{i})(5-\mathrm{i}) = 25+1 \quad \therefore \quad \underline{|uv| = \sqrt{26}}$

This confirms the general result that the 'modulus of a product is equal to the product of the moduli'. It can be derived from the above once we know that $(uv)^* = u^*v^*$, for any u and v, since it follows that $|uv|^2 = uu^*vv^* = |u|^2|v|^2$.

(d) $\left|\frac{u}{v}\right|^2 = \left(\frac{u}{v}\right)\left(\frac{u}{v}\right)^* = \left(\frac{-1+5i}{2}\right)\left(\frac{-1-5i}{2}\right) = \frac{1+25}{4} \qquad \therefore \ \left|\frac{u}{v}\right| = \sqrt{\frac{13}{2}}$

Again this provides a verification of the general result that 'the modulus of a quotient is equal to the ratio of the moduli'; and can be shown as such by using the fact that $(u/v)^* = u^*/v^*$ for any u and v.

(e) $\left|\frac{v}{u}\right|^2 = \left(\frac{v}{u}\right)\left(\frac{v}{u}\right)^* = \left(\frac{-1-5i}{13}\right)\left(\frac{-1+5i}{13}\right) = \frac{1+25}{169} \qquad \therefore \ \left|\frac{v}{u}\right| = \sqrt{\frac{2}{13}}$

This confirms the property that $|1/z| = 1/|z|$, for any complex number z; or, in this specific case, that $|v/u| = 1/|u/v|$.

7.3 If $z = 1 + i\sqrt{3}$, sketch the following in an Argand diagram: z, z^*, z^2, z^3, iz, and $1/z$.

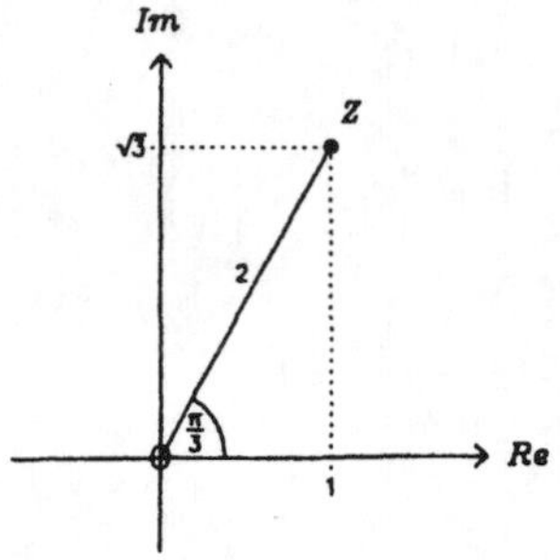

$$z = 1 + i\sqrt{3} = 2e^{i\pi/3}$$

That is to say, z has a modulus, or amplitude, of 2 and an argument, or phase, of 60°.

$$\therefore \quad z^* = 2e^{-i\pi/3}, \qquad z^2 = 4e^{i2\pi/3}, \qquad \text{and} \qquad z^3 = 8e^{i\pi} = -8$$

$$iz = e^{i\pi/2}\,2e^{i\pi/3} = 2e^{i(\pi/3+\pi/2)} = 2e^{i5\pi/6}, \qquad \text{and} \qquad \frac{1}{z} = \frac{1}{2}e^{-i\pi/3}$$

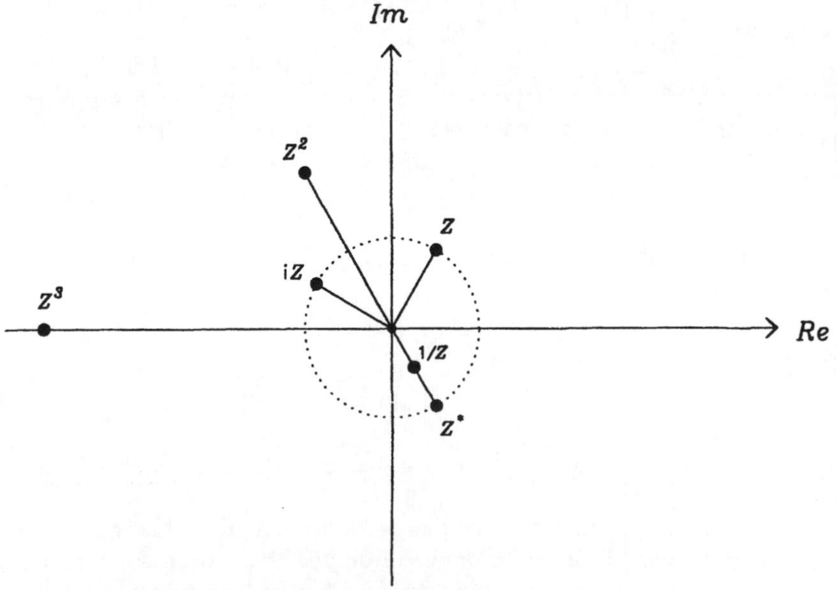

7.4 Solve the quadratic equation: $z^2 - z + 1 = 0$.

$$\text{If} \quad az^2 + bz + c = 0, \quad \text{then} \quad z = \frac{-b \pm \sqrt{b^2 - 4ac}}{2a}$$

$$z^2 - z + 1 = 0 \quad \Rightarrow \quad a = 1, \; b = -1 \text{ and } c = 1$$

$$\therefore \quad z = \frac{1 \pm \sqrt{1-4}}{2} = \underline{\frac{1 \pm i\sqrt{3}}{2}}$$

7.5 Solve the following equations:
(a) $z^5 = 1$, **(b)** $z^5 = 1 + i$, **(c)** $(z+1)^5 = 1$, and **(d)** $(z+1)^5 = z^5$.
Sketch the solutions for (a) on an Argand diagram.

(a) $\quad z^5 = 1 = e^{i2\pi n} \qquad$ where $n = 0, \pm 1, \pm 2, \pm 3, \ldots$

$$\therefore \quad \underline{z = e^{i2\pi n/5}} \qquad \text{for} \quad n = 0, 1, 2, 3, 4$$

Although the result for z holds for any integer n, only 5 distinct solutions emerge; an alternative choice for n which would specify them is $n = 0, \pm 1, \pm 2$. The appearance of five solutions was to be expected, as 'an nth-order polynomial has exactly n roots'; they are either real, or come as complex conjugate pairs (as in exercise 7.4).

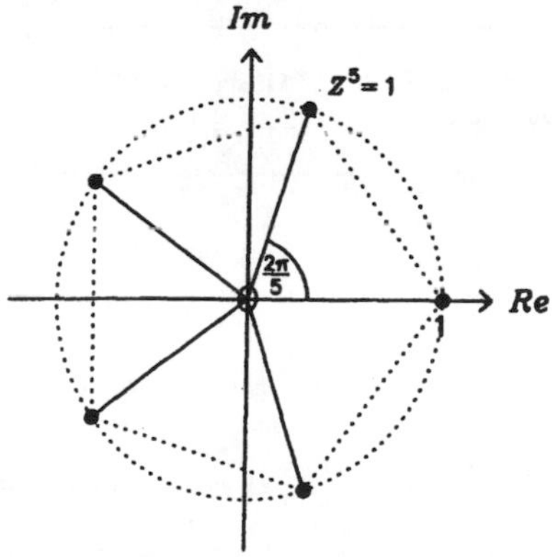

(b) $\quad z^5 = 1 + i = \sqrt{2}\, e^{i(\pi/4 + 2\pi n)} \qquad$ where $n = 0, \pm 1, \pm 2, \pm 3, \ldots$

$$\therefore \quad z = 2^{1/10}\, e^{i(\pi/4 + 2\pi n)/5} = \underline{2^{1/10}\, e^{i\pi(1+8n)/20}} \qquad \text{for } n = 0, 1, 2, 3, 4$$

(c) $(z+1)^5 = 1 = e^{i2\pi n}$ for $\pm$ integer n

$$\therefore \quad z+1 = e^{i2\pi n/5}$$

i.e. $\underline{z = e^{i2\pi n/5} - 1}$ for $n = 0, \pm 1, \pm 2$

(d) $\left(\frac{z+1}{z}\right)^5 = 1 = e^{i2\pi n}$ for $\pm$ integer n

$$\therefore \quad z+1 = z\,e^{i2\pi n/5}$$

$$\therefore \quad z\left(e^{i2\pi n/5} - 1\right) = 1$$

i.e. $\underline{z = \dfrac{1}{e^{i2\pi n/5} - 1}}$ for $n = 1, 2, 3, 4$

This is an interesting case because a seemingly fifth-order polynomial has only four roots: $n=0$ is not a permissible solution because the denominator, $e^{i2\pi n/5} - 1$ is then zero; the difficulty can also be seen two lines earlier, when $n=0$ yields $z+1=z$ and leads to the inconsistent conclusion that $1=0$! The resolution of the paradox is easily found in that a substitution of the binomial expansion of $(z+1)^5$ into the equation $(z+1)^5 = z^5$ shows that it reduces to a fourth-order polynomial: $5z^4 + 10z^3 + 10z^2 + 5z + 1 = 0$. The real and imaginary part of z can be obtained in the usual way by multiplying the top and bottom of the expression above by the complex conjugate of the denominator.

$(e^{i2\pi n/5} - 1)(e^{-i2\pi n/5} - 1)$
$= 2[1 - \cos(2\pi n/5)]$

> **7.6** By considering e^{iA} and e^{iB}, derive the compound-angle formulae for $\sin(A+B)$ and $\cos(A+B)$.

$$e^{i(A+B)} = e^{iA}\,e^{iB} \quad \text{and} \quad e^{i\theta} = \cos\theta + i\sin\theta$$

$$\therefore \quad \cos(A+B) + i\sin(A+B) = (\cos A + i\sin A)(\cos B + i\sin B)$$

$$= \cos A\cos B - \sin A\sin B$$

$$+ i(\sin A\cos B + \cos A\sin B)$$

Equating real parts $\Rightarrow$ $\underline{\cos(A+B) = \cos A\cos B - \sin A\sin B}$

Equating imaginary parts $\Rightarrow$ $\underline{\sin(A+B) = \sin A\cos B + \cos A\sin B}$

7.7 Use de Moivre's theorem to find expansions of $\sin 4\theta$ and $\cos 4\theta$ in terms of powers of $\sin\theta$ and $\cos\theta$.

$$\cos 4\theta + i\sin 4\theta = (\cos\theta + i\sin\theta)^4$$

$$= \cos^4\theta + 4i\cos^3\theta\sin\theta + 6i^2\cos^2\theta\sin^2\theta + 4i^3\cos\theta\sin^3\theta + i^4\sin^4\theta$$

$$= \cos^4\theta - 6\cos^2\theta\sin^2\theta + \sin^4\theta + i\left[4\cos^3\theta\sin\theta - 4\cos\theta\sin^3\theta\right]$$

Equating real parts $\Rightarrow$ $\underline{\cos 4\theta = \cos^4\theta - 6\cos^2\theta\sin^2\theta + \sin^4\theta}$

Equating imaginary parts $\Rightarrow$ $\underline{\sin 4\theta = 4\cos^3\theta\sin\theta - 4\cos\theta\sin^3\theta}$

This analysis with de Moivre's theorem is much more straightforward than the alternative procedure of repeatedly using the double-angle formulae for $\sin 2\theta$ and $\cos\theta$.

7.8 Show that $\cos^6\theta = (\cos 6\theta + 6\cos 4\theta + 15\cos 2\theta + 10)/32$.

$$\cos^6\theta = \left(\frac{e^{i\theta} + e^{-i\theta}}{2}\right)^6$$

$$= \frac{e^{i6\theta} + 6e^{i4\theta} + 15e^{i2\theta} + 20 + 15e^{-i2\theta} + 6e^{-i4\theta} + e^{-i6\theta}}{2^6}$$

$$= \frac{1}{32}\left[\frac{(e^{i6\theta} + e^{-i6\theta})}{2} + 6\frac{(e^{i4\theta} + e^{-i4\theta})}{2} + 15\frac{(e^{i2\theta} + e^{-i2\theta})}{2} + \frac{20}{2}\right]$$

$$= \underline{\frac{1}{32}(\cos 6\theta + 6\cos 4\theta + 15\cos 2\theta + 10)}$$

7.9 Use the definitions of hyperbolic functions to show that

$$\sinh(x+y) = \sinh x\cosh y + \cosh x\sinh y$$

and derive a similar expansion for $\cosh(x+y)$.

$$\sinh\theta = \frac{e^{\theta} - e^{-\theta}}{2} \quad \text{and} \quad \cosh\theta = \frac{e^{\theta} + e^{-\theta}}{2}$$

$$\sinh x\cosh y+\cosh x\sinh y=\left(\frac{e^x-e^{-x}}{2}\right)\left(\frac{e^y+e^{-y}}{2}\right)+\left(\frac{e^x+e^{-x}}{2}\right)\left(\frac{e^y-e^{-y}}{2}\right)$$

$$=\frac{e^{x+y}+e^{x-y}-e^{-x+y}-e^{-x-y}+e^{x+y}-e^{x-y}+e^{-x+y}-e^{-x-y}}{4}$$

$$=\frac{e^{x+y}-e^{-(x+y)}}{2}$$

$$=\underline{\sinh(x+y)}$$

$$\cosh(x+y)=\frac{e^{x+y}+e^{-(x+y)}}{2}$$

$$=\frac{2e^{x+y}+2e^{-x-y}+e^{-x+y}-e^{-x+y}+e^{x-y}-e^{x-y}}{4}$$

$$=\frac{e^{x+y}+e^{x-y}+e^{-x+y}+e^{-x-y}}{4}+\frac{e^{x+y}-e^{x-y}-e^{-x+y}+e^{-x-y}}{4}$$

$$=\left(\frac{e^x+e^{-x}}{2}\right)\left(\frac{e^y+e^{-y}}{2}\right)+\left(\frac{e^x-e^{-x}}{2}\right)\left(\frac{e^y-e^{-y}}{2}\right)$$

$$=\underline{\cosh x\cosh y+\sinh x\sinh y}$$

7.10 Find:

$$\textbf{(a)}\ \sum_{k=0}^{\infty}\frac{\cos k\theta}{k!}\quad\text{and}\quad\textbf{(b)}\ \int e^{ax}\sin(bx)\,dx$$

$$\textbf{(a)}\ \sum_{k=0}^{\infty}\frac{\cos k\theta}{k!}=\sum_{k=0}^{\infty}\frac{\mathcal{Re}\{e^{ik\theta}\}}{k!}=\sum_{k=0}^{\infty}\mathcal{Re}\left\{\frac{e^{ik\theta}}{k!}\right\}=\mathcal{Re}\left\{\sum_{k=0}^{\infty}\frac{(e^{i\theta})^k}{k!}\right\}$$

But $\displaystyle\sum_{k=0}^{\infty}\frac{\Phi^k}{k!}=1+\Phi+\frac{\Phi^2}{2!}+\frac{\Phi^3}{3!}+\cdots=\exp(\Phi)$

$$\therefore\ \sum_{k=0}^{\infty}\frac{\cos k\theta}{k!}=\mathcal{Re}\{\exp(e^{i\theta})\}$$

$$=\mathcal{Re}\{\exp(\cos\theta+i\sin\theta)\}$$

$$=\mathcal{Re}\{e^{\cos\theta}\,e^{i\sin\theta}\}$$

$$\text{i.e.} \quad \sum_{k=0}^{\infty} \frac{\cos k\theta}{k!} = \mathcal{R}e\left\{e^{\cos\theta}\left[\cos(\sin\theta) + \mathrm{i}\sin(\sin\theta)\right]\right\}$$

$$= \underline{e^{\cos\theta}\cos(\sin\theta)}$$

The corresponding sum of $\sin k\theta/k!$, from $k = 0$ to ∞, can be written straight down from the penultimate line, as being $e^{\cos\theta}\sin(\sin\theta)$, because the analysis is identical apart from the replacement of $\mathcal{R}e\{\}$ with $Im\{\}$.

$$\textbf{(b)} \quad \int e^{ax}\sin(bx)\,dx = \int e^{ax}\,Im\left\{e^{\mathrm{i}bx}\right\}dx$$

$$= \int Im\left\{e^{ax}e^{\mathrm{i}bx}\right\}dx$$

$$= Im\left\{\int e^{x(a+\mathrm{i}b)}\,dx\right\}$$

$$= Im\left\{\frac{e^{x(a+\mathrm{i}b)}}{a+\mathrm{i}b} + C\right\}$$

$$\text{But} \quad \frac{1}{a+\mathrm{i}b} = \left(\frac{1}{a+\mathrm{i}b}\right) \times \left(\frac{a-\mathrm{i}b}{a-\mathrm{i}b}\right) = \frac{a-\mathrm{i}b}{a^2+b^2}$$

$$\therefore \quad \int e^{ax}\sin(bx)\,dx = Im\left\{\frac{e^{ax}}{a^2+b^2}(a-\mathrm{i}b)(\cos bx + \mathrm{i}\sin bx) + C\right\}$$

$$= \underline{\frac{e^{ax}}{a^2+b^2}(a\sin bx - b\cos bx) + K}$$

Again, we can easily write down the integral of $e^{ax}\cos(bx)$ from the penultimate line by replacing $Im\{\}$ with $\mathcal{R}e\{\}$; so, we have really done two integrals for the price of one! We could have evaluated this integral by using integration-by-parts twice, but the complex number formulation accomplishes the same task with just one simple exponential integration. We should also note that the constants a and b were implicitly assumed to be real (as must be K).

8 Vectors

> **8.1** State which of the following quantities can be described by vectors: **(a)** temperature, **(b)** magnetic field, **(c)** acceleration, **(d)** force, **(e)** molecular weight, and **(f)** area.

(a) Temperature is always a <u>scalar</u> quantity because it makes no sense to talk about the direction of a temperature; it can vary with position, however, which is a vector. For example, a metal cube heated and cooled at its opposite corners would develop a temperature distribution $T(\underset{\sim}{r})$, a *scalar function* of the position vector $\underset{\sim}{r}$.

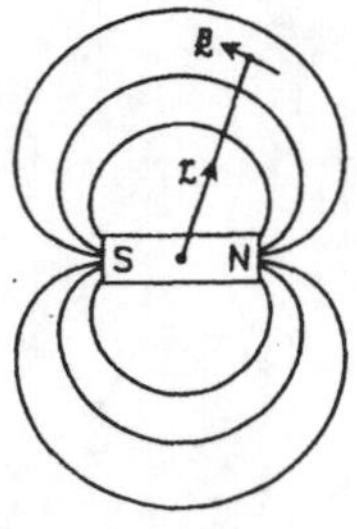

(b) Magnetic field is a <u>vector</u> quantity. A compass can be used to give its direction at any point. In general it is a *vector function* of position, written as $\underset{\sim}{B}(\underset{\sim}{r})$, since, as is illustrated by the field of a bar magnet, both the direction and magnitude of the field are usually functions of the position vector $\underset{\sim}{r}$.

$$\underset{\sim}{a} = \frac{d}{dt}(\underset{\sim}{v}) = \frac{d}{dt}\frac{d}{dt}(\underset{\sim}{r}) = \ddot{\underset{\sim}{r}}$$

$$\underset{\sim}{F} = \frac{d}{dt}(m\underset{\sim}{v}) = m\frac{d}{dt}(\underset{\sim}{v}) = m\underset{\sim}{a}$$

(c & d) Acceleration is the rate of change of velocity with time, and velocity is the rate of change of position with time. So, since position is given by the vector $\underset{\sim}{r}$, both velocity $\underset{\sim}{v} = d\underset{\sim}{r}/dt$ and acceleration $\underset{\sim}{a} = d\underset{\sim}{v}/dt$ must also be <u>vector</u> quantities. *Newton's Second Law of motion* states that force F is equal to the rate of change of momentum with time, and momentum is defined as the product of (scalar) mass m and (vector) velocity $\dot{\underset{\sim}{r}}$. If the mass of a body is constant, then the Second Law becomes the "$F = ma$" relation familiar from school physics, but with both force $\underset{\sim}{F}$ and acceleration $\underset{\sim}{a}$ as <u>vector</u> quantities.

(e) Molecular weight is a <u>scalar</u> quantity because it is the sum of the atomic weights of the atoms which comprise a molecule, and these weights are defined to be the mass of each atom relative to $1/12$ of the (scalar) mass of a ${}^{12}_{6}C$ atom. In other areas of science 'weight' is reserved for vector quantities. For example, the weight of a woman of mass m is the force she exerts on a set of bathroom scales; by Newton's Second Law this is given by $m\underset{\sim}{g}$, where $\underset{\sim}{g}$ is the vector acceleration due to gravity, which is directed towards the Earth's centre.

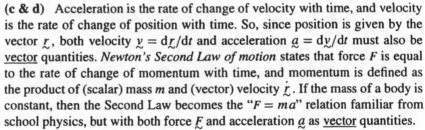

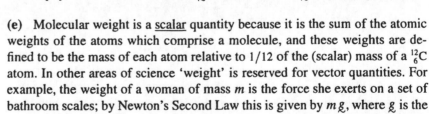

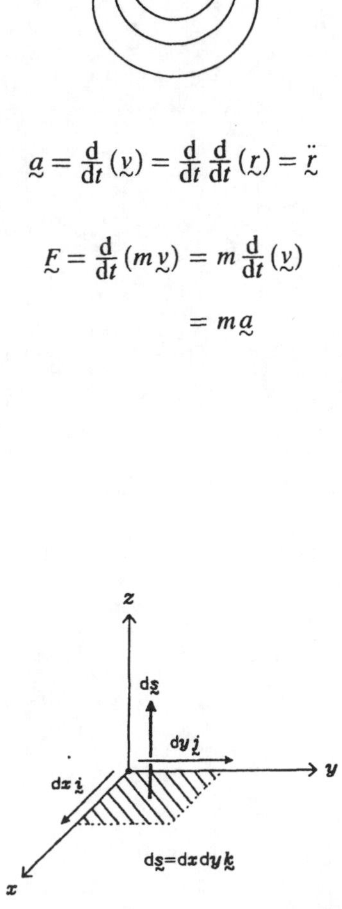

(f) Somewhat surprisingly, area can be considered as a <u>vector</u> quantity. For a flat-topped rectangular table with sides a and b, for example, it is defined to have a magnitude ab and a direction normal to the surface given by the "right-hand screw rule" (when rotating from a to b). Although most surfaces are far more complicated, any vanishingly-small portion can be treated as a plane and the total surface area obtained by the vector addition of all such segments of area that comprise the whole $\left(\underset{\sim}{s} = \int d\underset{\sim}{s}\right)$.

8.2 The four points A, B, C, and D respectively have position vectors

$$\underset{\sim}{a} = (1,2,3),\ \underset{\sim}{b} = (2,0,1),\ \underset{\sim}{c} = (1,1,1) \text{ and } \underset{\sim}{d} = (5,2,5)$$

Calculate: **(a)** $\underset{\sim}{a}+\underset{\sim}{b}-\underset{\sim}{c}-\underset{\sim}{d}$, **(b)** $2\underset{\sim}{a}-3\underset{\sim}{b}-5\underset{\sim}{c}+\underset{\sim}{d}/2$, and **(c)** the mid-points of $\overrightarrow{BC}$ and $\overrightarrow{AD}$.

(a)
$$\underset{\sim}{a}+\underset{\sim}{b}-\underset{\sim}{c}-\underset{\sim}{d} = (1+2-1-5\,,2+0-1-2\,,3+1-1-5) = \underline{(-3\,,-1\,,-2)}$$

(b)
$$2\underset{\sim}{a}-3\underset{\sim}{b}-5\underset{\sim}{c}+\underset{\sim}{d}/2 = [\,2(1)-3(2)-5(1)+5/2\,,\ 2(2)-3(0)-5(1)+2/2\,,\ 2(3)-3(1)-5(1)+5/2\,] = \underline{(-13/2\,,0\,,1/2)}$$

(c)
$$\text{Mid-point of } \overrightarrow{BC} = \underset{\sim}{b}+\tfrac{1}{2}(\underset{\sim}{c}-\underset{\sim}{b}) = \tfrac{1}{2}(\underset{\sim}{b}+\underset{\sim}{c}) = \underline{\tfrac{1}{2}(3\,,1\,,2)}$$

$$\text{Mid-point of } \overrightarrow{AD} = \tfrac{1}{2}(\underset{\sim}{a}+\underset{\sim}{d}) = \tfrac{1}{2}(6\,,4\,,8) = \underline{(3\,,2\,,4)}$$

8.3 Taking $\underset{\sim}{a}$, $\underset{\sim}{b}$, $\underset{\sim}{c}$ and $\underset{\sim}{d}$ from exercise 8.2, find **(a)** the vector equation of the line passing through A and C, **(b)** the vector equation of the line passing through the mid-points of $\overrightarrow{AB}$ and $\overrightarrow{CD}$, and **(c)** the Cartesian equations of the lines defined by $\underset{\sim}{r} = \underset{\sim}{a}+\lambda\underset{\sim}{b}$ and $\underset{\sim}{r} = \underset{\sim}{c}+\lambda\underset{\sim}{d}$.

(a) We need to get from the origin to a point on the line, say A, and then have the freedom to move an arbitrary scalar distance, given by the value of λ, along the direction $\overrightarrow{AC} = \underset{\sim}{c}-\underset{\sim}{a}$. This gives

$$\underset{\sim}{r} = \underset{\sim}{a}+\lambda(\underset{\sim}{c}-\underset{\sim}{a}) = (1\,,2\,,3)+\lambda(1-1\,,1-2\,,1-3) = \underline{(1\,,2\,,3)-\lambda(0\,,1\,,2)}$$

(b)
$$\text{Mid-point of } \overrightarrow{AB},\ \underset{\sim}{e} = \tfrac{1}{2}(\underset{\sim}{a}+\underset{\sim}{b}) = \tfrac{1}{2}(3\,,2\,,4)$$

$$\text{Mid-point of } \overrightarrow{CD},\ \underset{\sim}{f} = \tfrac{1}{2}(\underset{\sim}{c}+\underset{\sim}{d}) = \tfrac{1}{2}(6\,,3\,,6)$$

$$\therefore \text{ Equation of line is } \underset{\sim}{r} = \underset{\sim}{e}+\lambda(\underset{\sim}{f}-\underset{\sim}{e}) = \underline{\tfrac{1}{2}(3\,,2\,,4)+\tfrac{\lambda}{2}(3\,,1\,,2)}$$

(c) $\underset{\sim}{r} = \underset{\sim}{a} + \lambda \underset{\sim}{b} \;\Rightarrow\; (x, y, z) = (1+2\lambda, 2, 3+\lambda)$

$$\therefore\; \lambda = \frac{x-1}{2} = z-3 \text{ and } y = 2$$

$$\underset{\sim}{r} = \underset{\sim}{c} + \lambda \underset{\sim}{d} \;\Rightarrow\; (x, y, z) = (1+5\lambda, 1+2\lambda, 1+5\lambda)$$

$$\therefore\; \lambda = \frac{x-1}{5} = \frac{y-1}{2} = \frac{z-1}{5}$$

> **8.4** Taking $\underset{\sim}{a}$, $\underset{\sim}{b}$, $\underset{\sim}{c}$ and $\underset{\sim}{d}$ from exercise 8.2, find $\underset{\sim}{a}\cdot\underset{\sim}{b}$, $\underset{\sim}{a}\cdot\underset{\sim}{c}$ and $\underset{\sim}{a}\cdot\underset{\sim}{d}$. Find the angles between $\underset{\sim}{b}$ and $\underset{\sim}{c}$, and between $\underset{\sim}{c}$ and $\underset{\sim}{d}$. Evaluate $(\underset{\sim}{a}\cdot\underset{\sim}{c})\,\underset{\sim}{b}$ and $(\underset{\sim}{a}\cdot\underset{\sim}{b})\,\underset{\sim}{c}$.

$$\underset{\sim}{a}\cdot\underset{\sim}{b} = (1,2,3)\cdot(2,0,1) = 1\times 2 + 2\times 0 + 3\times 1 = \underline{5}$$

$$\underset{\sim}{a}\cdot\underset{\sim}{c} = (1,2,3)\cdot(1,1,1) = 1\times 1 + 2\times 1 + 3\times 1 = \underline{6}$$

$$\underset{\sim}{a}\cdot\underset{\sim}{d} = (1,2,3)\cdot(5,2,5) = 1\times 5 + 2\times 2 + 3\times 5 = \underline{24}$$

$$\cos(B\hat{O}C) = \frac{\underset{\sim}{b}\cdot\underset{\sim}{c}}{|\underset{\sim}{b}|\,|\underset{\sim}{c}|} = \frac{2\times 1 + 0\times 1 + 1\times 1}{\sqrt{2^2+1^2}\,\sqrt{1^2+1^2+1^2}} = \frac{3}{\sqrt{5}\,\sqrt{3}}$$

$$\therefore\; B\hat{O}C = \cos^{-1}\sqrt{3/5} = \underline{39.2^\circ}$$

$$C\hat{O}D = \cos^{-1}\left(\frac{\underset{\sim}{c}\cdot\underset{\sim}{d}}{|\underset{\sim}{c}|\,|\underset{\sim}{d}|}\right) = \cos^{-1}\left(\frac{12}{\sqrt{3}\,\sqrt{54}}\right) = \cos^{-1}\sqrt{8/9} = \underline{19.5^\circ}$$

$$(\underset{\sim}{a}\cdot\underset{\sim}{c})\,\underset{\sim}{b} = 6(2,0,1) = \underline{(12,0,6)} \quad\text{and}\quad (\underset{\sim}{a}\cdot\underset{\sim}{b})\,\underset{\sim}{c} = \underline{(5,5,5)}$$

We should emphasise that terms like $(\underset{\sim}{a}\cdot\underset{\sim}{c})\,\underset{\sim}{b}$ are vectors, and are best thought of as the scalar '$\underset{\sim}{a}$ dot $\underset{\sim}{c}$' *lots of* the vector $\underset{\sim}{b}$.

> **8.5** Use the dot product to show $\cos(A \pm B) = \cos A \cos B \mp \sin A \sin B$.

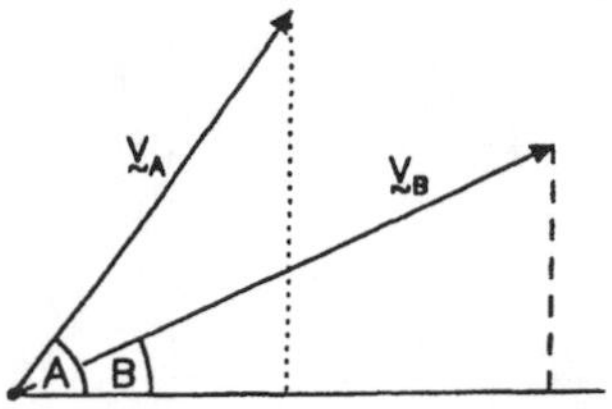

Consider two vectors $\underset{\sim}{v}_A$ and $\underset{\sim}{v}_B$ in the x-y ($z = 0$) plane which make angles of A and B with the x-axis respectively. Then

$$\underset{\sim}{v}_A = |\underset{\sim}{v}_A|(\cos A, \sin A, 0) \quad\text{and}\quad \underset{\sim}{v}_B = |\underset{\sim}{v}_B|(\cos B, \sin B, 0)$$

$$\therefore\; \underset{\sim}{v}_A\cdot\underset{\sim}{v}_B = |\underset{\sim}{v}_A|\,|\underset{\sim}{v}_B|\cos(A-B)$$

$$\text{and}\;\; \underset{\sim}{v}_A\cdot\underset{\sim}{v}_B = |\underset{\sim}{v}_A|\,|\underset{\sim}{v}_B|(\cos A\cos B + \sin A\sin B + 0)$$

i.e. $\underline{\cos(A-B) = \cos A \cos B + \sin A \sin B}$

Putting $B = -C \quad \Rightarrow \quad \underline{\cos(A+C) = \cos A \cos C - \sin A \sin C}$

$\sin(-\theta) = -\sin(\theta)$

> **8.6** Taking $\underset{\sim}{a}$, $\underset{\sim}{b}$, $\underset{\sim}{c}$ and $\underset{\sim}{d}$ from exercise 8.2, find $\underset{\sim}{a}\times\underset{\sim}{b}$, $\underset{\sim}{a}\times\underset{\sim}{c}$, and $\underset{\sim}{a}\times\underset{\sim}{d}$. Find the angles between $\underset{\sim}{b}$ and $\underset{\sim}{c}$, and between $\underset{\sim}{c}$ and $\underset{\sim}{d}$. Express the two lines of exercise 8.3(c) in the form $\underset{\sim}{r}\times\underset{\sim}{p} = \underset{\sim}{q}$.

$$\underset{\sim}{a}\times\underset{\sim}{b} = (1,2,3)\times(2,0,1)$$
$$= (2\times1-3\times0,\,3\times2-1\times1,\,1\times0-2\times2) = \underline{(2,5,-4)}$$

$$\underset{\sim}{a}\times\underset{\sim}{c} = (1,2,3)\times(1,1,1)$$
$$= (2\times1-3\times1,\,3\times1-1\times1,\,1\times1-2\times1) = \underline{(-1,2,-1)}$$

$$\underset{\sim}{a}\times\underset{\sim}{d} = (1,2,3)\times(5,2,5)$$
$$= (2\times5-3\times2,\,3\times5-1\times5,\,1\times2-2\times5) = \underline{(4,10,-8)}$$

$$\sin(B\hat{O}C) = \frac{|\underset{\sim}{b}\times\underset{\sim}{c}|}{|\underset{\sim}{b}|\,|\underset{\sim}{c}|} = \frac{|(-1,-1,2)|}{\sqrt{2^2+1^2}\,\sqrt{1^2+1^2+1^2}} = \frac{\sqrt{6}}{\sqrt{5}\,\sqrt{3}}$$

$$\therefore\quad B\hat{O}C = \sin^{-1}\sqrt{2/5} = \underline{39.2^\circ} \quad \text{(as before)}$$

$$C\hat{O}D = \sin^{-1}\left(\frac{|\underset{\sim}{c}\times\underset{\sim}{d}|}{|\underset{\sim}{c}|\,|\underset{\sim}{d}|}\right) = \sin^{-1}\left(\frac{\sqrt{18}}{\sqrt{3}\,\sqrt{54}}\right) = \sin^{-1}(1/3) = \underline{19.5^\circ}$$

Since the vector product of any vector with itself (as in $\underset{\sim}{b}\times\underset{\sim}{b}$) is always zero, the equation $\underset{\sim}{r} = \underset{\sim}{a} + \lambda\underset{\sim}{b}$ can be expressed in a form which excludes λ by taking the vector product of both sides with $\underset{\sim}{b}$.

$$\underset{\sim}{r}\times\underset{\sim}{b} = \underset{\sim}{a}\times\underset{\sim}{b} + \lambda\underset{\sim}{b}\times\underset{\sim}{b} = \underset{\sim}{a}\times\underset{\sim}{b} \quad\Rightarrow\quad \underline{\underset{\sim}{r}\times(2,0,1) = (2,5,-4)}$$

$$\text{Similarly}\quad \underset{\sim}{r}\times\underset{\sim}{d} = \underset{\sim}{c}\times\underset{\sim}{d} \quad\Rightarrow\quad \underline{\underset{\sim}{r}\times(5,2,5) = (3,0,-3)}$$

> **8.7** Use the cross product to obtain the 'sine rule' for triangles.

Consider the triangle formed by three vectors $\underset{\sim}{a}$, $\underset{\sim}{b}$ and $\underset{\sim}{c}$ such that $\underset{\sim}{c} = \underset{\sim}{a} + \underset{\sim}{b}$.

$$\therefore\quad \underset{\sim}{a}\times\underset{\sim}{c} = \underset{\sim}{a}\times(\underset{\sim}{a}+\underset{\sim}{b}) = \underset{\sim}{a}\times\underset{\sim}{a} + \underset{\sim}{a}\times\underset{\sim}{b} = \underset{\sim}{a}\times\underset{\sim}{b}$$

$\underset{\sim}{a}\times\underset{\sim}{a} = \underset{\sim}{b}\times\underset{\sim}{b} = 0$

Defining β as the angle between $\underset{\sim}{a}$ and $\underset{\sim}{c}$, and hence the angle opposite to the vector side $\underset{\sim}{b}$, and θ as the angle between $\underset{\sim}{a}$ and $\underset{\sim}{b}$, we have

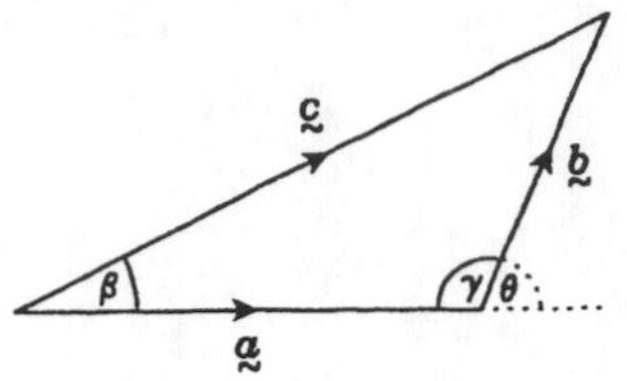

$$\left.\begin{aligned} \underset{\sim}{a}\times\underset{\sim}{b} &= |\underset{\sim}{a}|\,|\underset{\sim}{b}|\sin\theta \\ \text{and}\quad \underset{\sim}{a}\times\underset{\sim}{b} = \underset{\sim}{a}\times\underset{\sim}{c} &= |\underset{\sim}{a}|\,|\underset{\sim}{c}|\sin\beta \end{aligned}\right\} \quad \therefore\quad |\underset{\sim}{b}|\sin\theta = |\underset{\sim}{c}|\sin\beta$$

If γ is the angle opposite vector side $\underset{\sim}{c}$, then the sum of γ and θ must be π radians, or 180°, and $\sin\gamma = \sin(\pi - \theta) = \sin\theta$.

$\alpha + \beta + \gamma = \pi$

$$\text{i.e.}\quad \underline{\frac{|\underset{\sim}{b}|}{\sin\beta} = \frac{|\underset{\sim}{c}|}{\sin\gamma}}\,;\quad \text{and, similarly, } \underset{\sim}{b}\times\underset{\sim}{c} \;\Rightarrow\; \underline{\frac{|\underset{\sim}{a}|}{\sin\alpha} = \frac{|\underset{\sim}{c}|}{\sin\gamma}}$$

8.8 Taking $\underset{\sim}{a}$, $\underset{\sim}{b}$, $\underset{\sim}{c}$ and $\underset{\sim}{d}$ from exercise 8.2, find $\underset{\sim}{a}\cdot(\underset{\sim}{b}\times\underset{\sim}{c})$, $\underset{\sim}{a}\cdot(\underset{\sim}{c}\times\underset{\sim}{d})$ and $\underset{\sim}{a}\cdot(\underset{\sim}{b}\times\underset{\sim}{d})$. Which three position vectors are coplanar? Find an equation for this plane in its Cartesian and vector forms, and its perpendicular distance from the origin.

$$\begin{aligned} \underset{\sim}{a}\cdot(\underset{\sim}{b}\times\underset{\sim}{c}) &= (1,2,3)\cdot[(2,0,1)\times(1,1,1)] \\ &= (1,2,3)\cdot(-1,-1,2) = \underline{3} \end{aligned}$$

$$\begin{aligned} \underset{\sim}{a}\cdot(\underset{\sim}{c}\times\underset{\sim}{d}) &= (1,2,3)\cdot[(1,1,1)\times(5,2,5)] \\ &= (1,2,3)\cdot(3,0,-3) = \underline{-6} \end{aligned}$$

$$\begin{aligned} \underset{\sim}{a}\cdot(\underset{\sim}{b}\times\underset{\sim}{d}) &= (1,2,3)\cdot[(2,0,1)\times(5,2,5)] \\ &= (1,2,3)\cdot(-2,-5,4) = \underline{0} \end{aligned}$$

If the scalar triple product of any three vectors is zero, then the parallelepiped formed by these vectors has zero volume. This can only be true if the three vectors are confined to a plane or a single line. In this case, $\underset{\sim}{a}\cdot(\underset{\sim}{b}\times\underset{\sim}{d}) = 0$ shows that the position vectors $\underset{\sim}{a}$, $\underset{\sim}{b}$ and $\underset{\sim}{d}$ are coplanar.

To find a normal, $\underset{\sim}{n}$, to this plane we take the vector product of any two of the coplanar vectors, say $\underset{\sim}{a}\times\underset{\sim}{b}$:

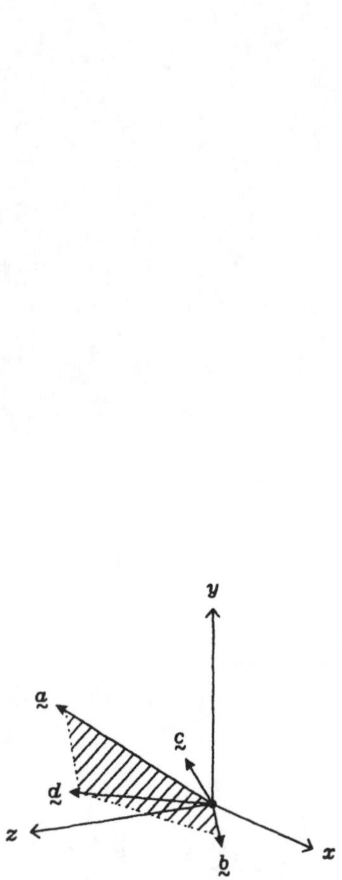

$$\underset{\sim}{n} = \underset{\sim}{a}\times\underset{\sim}{b} = (2,5,-4) \qquad \therefore\quad \text{Eqn of plane is}\quad \underset{\sim}{r}\cdot(2,5,-4) = d$$

where $\underset{\sim}{r}$ is a point in the plane and d is a (scalar) constant. Taking $\underset{\sim}{r} = \underset{\sim}{a}$, for example, we find that $d = (1,2,3)\cdot(2,5,-4) = 0$. Hence

$$\underline{\underset{\sim}{r}\cdot(2,5,-4) = 0} \qquad \text{or} \qquad (x,y,z)\cdot(2,5,-4) = \underline{2x + 5y - 4z = 0}$$

The perpendicular distance of the origin to the plane is <u>zero</u> because it is given by the right-hand side of the vector equation for a plane when the normal vector has unit length. This can be achieved by dividing both sides of the equation by $|(2,5,-4)|$ which, of course, still leaves zero on the right-hand side.

8.9 Find the scalar triple product of the vectors $(1,2,4)$, $(2,0,-3)$ and $(-4,4,17)$. Are they linearly independent? Can the third vector be expressed as a linear combination of the first two? If so, find this linear combination.

$$(1,2,4)\cdot[(2,0,-3)\times(-4,4,17)] = (1,2,4)\cdot(12,-22,8) = \underline{0}$$

Since the scalar triple product of these three vectors is zero, we know that they must either be confined to a plane or lie along a single line. In such cases we say that the three vectors are linearly dependent, or <u>not</u> linearly independent. The vectors in question are not simple scalar multiples of each other, and so are not mutually parallel. They must therefore lie in a plane, so that the third vector <u>can</u> be expressed as a *linear combination* of the first two vectors:

$$(-4,4,17) = \alpha(1,2,4) + \beta(2,0,-3)$$

where α and β are constants. These coefficients can be evaluated by dotting both sides of the equation with $(3,0,2)$, a vector chosen to be perpendicular to $(2,0,-3)$.

$$(-4,4,17)\cdot(3,0,2) = \alpha(1,2,4)\cdot(3,0,2) + \beta(2,0,-3)\cdot(3,0,2)$$

$$\therefore\quad 22 = 11\alpha + 0 \quad\Rightarrow\quad \alpha = 2$$

Substituting $\alpha = 2$, and equating the x (or z) components, yields $\beta = -3$.

$$\text{i.e.}\quad \underline{(-4,4,17) = 2(1,2,4) - 3(2,0,-3)}$$

8.10 Use the results of exercise 8.4 to verify the 'ABACAB' identity.

$$\underset{\sim}{a}\times(\underset{\sim}{b}\times\underset{\sim}{c}) = (1,2,3)\times[(2,0,1)\times(1,1,1)]$$
$$= (1,2,3)\times(-1,-1,2) = \underline{(7,-5,1)}$$

$$(\underset{\sim}{a}\cdot\underset{\sim}{c})\underset{\sim}{b} - (\underset{\sim}{a}\cdot\underset{\sim}{b})\underset{\sim}{c} = 6(2,0,1) - 5(1,1,1) = \underline{(7,-5,1)}$$

Hence the 'ABACAB' identity, $\underbrace{\underset{\sim}{a}\times(\underset{\sim}{b}}_{AB}\times\underset{\sim}{c}) = \underbrace{(\underset{\sim}{a}\cdot\underset{\sim}{c})}_{AC}\underset{\sim}{b} - \underbrace{(\underset{\sim}{a}\cdot\underset{\sim}{b})}_{AB}\underset{\sim}{c}$, is satisfied.

8.11 The *reciprocal vectors* of the set $\underset{\sim}{a}$, $\underset{\sim}{b}$ and $\underset{\sim}{c}$ are defined by

$$\underset{\sim}{a}' = (\underset{\sim}{b} \times \underset{\sim}{c})/s\,, \quad \underset{\sim}{b}' = (\underset{\sim}{c} \times \underset{\sim}{a})/s \quad \text{and} \quad \underset{\sim}{c}' = (\underset{\sim}{a} \times \underset{\sim}{b})/s$$

where $s = \underset{\sim}{a} \cdot (\underset{\sim}{b} \times \underset{\sim}{c})$. Show that $\underset{\sim}{a} \cdot \underset{\sim}{a}' = \underset{\sim}{b} \cdot \underset{\sim}{b}' = \underset{\sim}{c} \cdot \underset{\sim}{c}' = 1$, and that $\underset{\sim}{a} \cdot \underset{\sim}{b}' = \underset{\sim}{a} \cdot \underset{\sim}{c}' = 0$; evaluate the scalar triple product of the reciprocal vectors in terms of s. If $\underset{\sim}{x}$ is expressed as a linear combination of the reciprocal vectors, show that the coefficient of $\underset{\sim}{a}'$ is $\underset{\sim}{a} \cdot \underset{\sim}{x}$.

Dotting $\underset{\sim}{a}$, $\underset{\sim}{b}$ and $\underset{\sim}{c}$ with $\underset{\sim}{a}'$, $\underset{\sim}{b}'$ and $\underset{\sim}{c}'$ respectively, gives

$$[\underset{\sim}{a}, \underset{\sim}{b}, \underset{\sim}{c}] = \underset{\sim}{a} \cdot (\underset{\sim}{b} \times \underset{\sim}{c}) = \underset{\sim}{b} \cdot (\underset{\sim}{c} \times \underset{\sim}{a}) = \underset{\sim}{c} \cdot (\underset{\sim}{a} \times \underset{\sim}{b})$$

$$\underset{\sim}{a} \cdot \underset{\sim}{a}' = \underset{\sim}{a} \cdot (\underset{\sim}{b} \times \underset{\sim}{c})/s = s/s = 1$$

$$\underset{\sim}{b} \cdot \underset{\sim}{b}' = \underset{\sim}{b} \cdot (\underset{\sim}{c} \times \underset{\sim}{a})/s = s/s = 1$$

$$\underset{\sim}{c} \cdot \underset{\sim}{c}' = \underset{\sim}{c} \cdot (\underset{\sim}{a} \times \underset{\sim}{b})/s = s/s = 1$$

Whereas, dotting $\underset{\sim}{a}$ with $\underset{\sim}{b}'$ or $\underset{\sim}{c}'$ yields

$$\underset{\sim}{a} \cdot \underset{\sim}{b}' = \underset{\sim}{a} \cdot (\underset{\sim}{c} \times \underset{\sim}{a})/s = 0 \quad \text{and} \quad \underset{\sim}{a} \cdot \underset{\sim}{c}' = \underset{\sim}{a} \cdot (\underset{\sim}{a} \times \underset{\sim}{b})/s = 0$$

because the scalar triple products contain two identical vectors.

The scalar triple product of the reciprocal set is

$$\begin{aligned}
\underset{\sim}{a}' \cdot (\underset{\sim}{b}' \times \underset{\sim}{c}') &= (\underset{\sim}{b} \times \underset{\sim}{c}) \cdot \big((\underset{\sim}{c} \times \underset{\sim}{a}) \times (\underset{\sim}{a} \times \underset{\sim}{b})\big)/s^3 \\
&= (\underset{\sim}{b} \times \underset{\sim}{c}) \cdot \Big([\underset{\sim}{b} \cdot (\underset{\sim}{c} \times \underset{\sim}{a})]\underset{\sim}{a} - [\underset{\sim}{a} \cdot (\underset{\sim}{c} \times \underset{\sim}{a})]\underset{\sim}{b}\Big)/s^3 \\
&= (\underset{\sim}{b} \times \underset{\sim}{c}) \cdot (s\underset{\sim}{a} - 0)/s^3 \\
&= \underset{\sim}{a} \cdot (\underset{\sim}{b} \times \underset{\sim}{c})/s^2 \\
&= s/s^2 = \underline{1/s}
\end{aligned}$$

If $\underset{\sim}{x} = \alpha\, \underset{\sim}{a}' + \beta\, \underset{\sim}{b}' + \gamma\, \underset{\sim}{c}'$, then

$$\underset{\sim}{a} \cdot \underset{\sim}{x} = \alpha\, \underset{\sim}{a} \cdot \underset{\sim}{a}' + \beta\, \underset{\sim}{a} \cdot \underset{\sim}{b}' + \gamma\, \underset{\sim}{a} \cdot \underset{\sim}{c}' = \alpha + 0 + 0$$

i.e. $\underline{\alpha = \underset{\sim}{a} \cdot \underset{\sim}{x}}$; and, similarly, $\underline{\beta = \underset{\sim}{b} \cdot \underset{\sim}{x}}$ and $\underline{\gamma = \underset{\sim}{c} \cdot \underset{\sim}{x}}$

Reciprocal vectors are often used to simplify problems in crystallography and solid-state physics.

9 Matrices

9.1 Find $\underset{\approx}{A}+\underset{\approx}{B}$, $\underset{\approx}{A}-\underset{\approx}{B}$, $\underset{\approx}{A}\,\underset{\approx}{B}$, and $\underset{\approx}{B}\,\underset{\approx}{A}$ when

$$\underset{\approx}{A}=\begin{pmatrix}2&1\\1&2\end{pmatrix} \quad \text{and} \quad \underset{\approx}{B}=\begin{pmatrix}3&3\\0&4\end{pmatrix}$$

Verify that $(\underset{\approx}{A}\,\underset{\approx}{B})^T=\underset{\approx}{B}^T\underset{\approx}{A}^T$, and that $\det(\underset{\approx}{A}\,\underset{\approx}{B})=\det(\underset{\approx}{A})\det(\underset{\approx}{B})$.

$$\underset{\approx}{A}+\underset{\approx}{B} = \begin{pmatrix}2+3&1+3\\1+0&2+4\end{pmatrix}=\underline{\begin{pmatrix}5&4\\1&6\end{pmatrix}}$$

$$\underset{\approx}{A}-\underset{\approx}{B} = \begin{pmatrix}2-3&1-3\\1-0&2-4\end{pmatrix}=\underline{\begin{pmatrix}-1&-2\\1&-2\end{pmatrix}}$$

$$\underset{\approx}{A}\,\underset{\approx}{B} = \begin{pmatrix}2\times3+1\times0&2\times3+1\times4\\1\times3+2\times0&1\times3+2\times4\end{pmatrix}=\underline{\begin{pmatrix}6&10\\3&11\end{pmatrix}}$$

$$\underset{\approx}{B}\,\underset{\approx}{A} = \begin{pmatrix}3\times2+3\times1&3\times1+3\times2\\0\times2+4\times1&0\times1+4\times2\end{pmatrix}=\underline{\begin{pmatrix}9&9\\4&8\end{pmatrix}}$$

This simple example illustrates that matrix multiplication is not commutative: $\underset{\approx}{A}\,\underset{\approx}{B}\neq\underset{\approx}{B}\,\underset{\approx}{A}$ in general, even when both products exist (or are permissible).

$$\underset{\approx}{B}^T\underset{\approx}{A}^T=\begin{pmatrix}3&0\\3&4\end{pmatrix}\begin{pmatrix}2&1\\1&2\end{pmatrix}=\begin{pmatrix}6&3\\10&11\end{pmatrix}=\begin{pmatrix}6&10\\3&11\end{pmatrix}^T=(\underset{\approx}{A}\,\underset{\approx}{B})^T$$

$$\det(\underset{\approx}{A}) = \begin{vmatrix}2&1\\1&2\end{vmatrix}=2\times2-1\times1=3$$

$$\det(\underset{\approx}{B}) = \begin{vmatrix}3&3\\0&4\end{vmatrix}=3\times4-0\times3=12$$

$$\det(\underset{\approx}{A}\,\underset{\approx}{B}) = \begin{vmatrix}6&10\\3&11\end{vmatrix}=6\times11-3\times10=36=\det(\underset{\approx}{A})\det(\underset{\approx}{B})$$

9.2 Show how the determinant of a 3×3 matrix can be used to evaluate both a cross product and a scalar triple product.

Let the vectors $\underset{\sim}{a}$, $\underset{\sim}{b}$ and $\underset{\sim}{c}$ have components

$$\underset{\sim}{a} = (a_1, a_2, a_3) = a_1 \underset{\sim}{i} + a_2 \underset{\sim}{j} + a_3 \underset{\sim}{k}$$

$$\underset{\sim}{b} = (b_1, b_2, b_3) = b_1 \underset{\sim}{i} + b_2 \underset{\sim}{j} + b_3 \underset{\sim}{k}$$

$$\underset{\sim}{c} = (c_1, c_2, c_3) = c_1 \underset{\sim}{i} + c_2 \underset{\sim}{j} + c_3 \underset{\sim}{k}$$

where $\underset{\sim}{i}$, $\underset{\sim}{j}$ and $\underset{\sim}{k}$ are unit vectors in the x, y and z directions respectively. Then,

$$\begin{vmatrix} \underset{\sim}{i} & \underset{\sim}{j} & \underset{\sim}{k} \\ a_1 & a_2 & a_3 \\ b_1 & b_2 & b_3 \end{vmatrix} = (a_2 b_3 - b_2 a_3)\, \underset{\sim}{i} - (a_1 b_3 - b_1 a_3)\, \underset{\sim}{j} + (a_1 b_2 - b_1 a_2) \underset{\sim}{k}$$

$$= (a_2 b_3 - b_2 a_3\,,\; a_3 b_1 - b_3 a_1\,,\; a_1 b_2 - b_1 a_2) \;=\; \underline{\underset{\sim}{a} \times \underset{\sim}{b}}$$

$$\begin{vmatrix} c_1 & c_2 & c_3 \\ a_1 & a_2 & a_3 \\ b_1 & b_2 & b_3 \end{vmatrix} = (a_2 b_3 - b_2 a_3)\, c_1 + (a_3 b_1 - b_3 a_1)\, c_2 + (a_1 b_2 - b_1 a_2)\, c_3$$

$$= (\underset{\sim}{a} \times \underset{\sim}{b})_1\, c_1 + (\underset{\sim}{a} \times \underset{\sim}{b})_2\, c_2 + (\underset{\sim}{a} \times \underset{\sim}{b})_3\, c_3 \;=\; \underline{(\underset{\sim}{a} \times \underset{\sim}{b}) \cdot \underset{\sim}{c}}$$

The general properties of determinants can be used to infer some of those of a cross and scalar triple product:

(i) $\underset{\sim}{a} \times \underset{\sim}{b} = -\underset{\sim}{b} \times \underset{\sim}{a}$, because the determinant is multiplied by -1 if two rows (or columns) are interchanged;

(ii) $\underset{\sim}{a} \times \underset{\sim}{b} = 0$ if $\underset{\sim}{a}$ is parallel to $\underset{\sim}{b}$, because a determinant is zero if two rows are the same;

(iii) $(\underset{\sim}{a} \times \underset{\sim}{b}) \cdot \underset{\sim}{c} = 0$ if $\underset{\sim}{a}$, $\underset{\sim}{b}$, and $\underset{\sim}{c}$ are not linearly independent, so that they lie in a plane (or along a line), for then any row minus a suitable combination of the other two will result in a whole line of zeros.

9.3 Find the inverse of the following matrix

$$\underset{\approx}{C} = \begin{pmatrix} 2 & -1 & 1 \\ 1 & -1 & 2 \\ -1 & 1 & -1 \end{pmatrix}$$

and verify that $\underset{\approx}{C}\,\underset{\approx}{C}^{-1} = \underset{\approx}{C}^{-1}\underset{\approx}{C} = \underset{\approx}{I}$.

$$\underset{\approx}{C}^T = \begin{pmatrix} 2 & 1 & -1 \\ -1 & -1 & 1 \\ 1 & 2 & -1 \end{pmatrix}$$

$$\text{adj}(\underset{\approx}{C}) = \text{matrix of cofactors of } \underset{\approx}{C}^T$$

$$= \begin{pmatrix} (-1)\times(-1) - 2\times 1 & 1\times 1 - (-1)\times(-1) & (-1)\times 2 - 1\times(-1) \\ 2\times(-1) - 1\times(-1) & 2\times(-1) - 1\times(-1) & 1\times 1 - 2\times 2 \\ 1\times 1 - (-1)\times(-1) & (-1)\times(-1) - 2\times 1 & 2\times(-1) - (-1)\times 1 \end{pmatrix}$$

$$= \begin{pmatrix} -1 & 0 & -1 \\ -1 & -1 & -3 \\ 0 & -1 & -1 \end{pmatrix}$$

$$\det(\underset{\approx}{C}) = \det(\underset{\approx}{C}^T)$$

$$= \text{dot product of any row, or column, with its cofactors}$$

$$= 2\times(-1) + 1\times 0 + (-1)\times(-1) = -1$$

$$\therefore\quad \underset{\approx}{C}^{-1} = \frac{\text{adj}(\underset{\approx}{C})}{\det(\underset{\approx}{C})} = \underline{\begin{pmatrix} 1 & 0 & 1 \\ 1 & 1 & 3 \\ 0 & 1 & 1 \end{pmatrix}}$$

$$\underset{\approx}{C}\,\underset{\approx}{C}^{-1} = \begin{pmatrix} 2-1+0 & 0-1+1 & 2-3+1 \\ 1-1+0 & 0-1+2 & 1-3+2 \\ -1+1+0 & 0+1-1 & -1+3-1 \end{pmatrix} = \begin{pmatrix} 1 & 0 & 0 \\ 0 & 1 & 0 \\ 0 & 0 & 1 \end{pmatrix} = \underset{\approx}{I}$$

$$\underset{\approx}{C}^{-1}\underset{\approx}{C} = \begin{pmatrix} 2+0-1 & -1+0+1 & 1+0-1 \\ 2+1-3 & -1-1+3 & 1+2-3 \\ 0+1-1 & 0-1+1 & 0+2-1 \end{pmatrix} = \begin{pmatrix} 1 & 0 & 0 \\ 0 & 1 & 0 \\ 0 & 0 & 1 \end{pmatrix} = \underset{\approx}{I}$$

9.4 Show that the eigenvalues of an Hermitian matrix are real, and that the eigenvectors corresponding to distinct eigenvalues are orthogonal.

The equations for the jth and k^{th} eigenvalues, λ_j and λ_k, and the corresponding eigenvectors, $\underset{\sim}{x}_j$ and $\underset{\sim}{x}_k$, of a matrix $\underset{\approx}{A}$ are

$$\underset{\approx}{A}\,\underset{\sim}{x}_j = \lambda_j\,\underset{\sim}{x}_j \quad \text{— (1)} \qquad \text{and} \qquad \underset{\approx}{A}\,\underset{\sim}{x}_k = \lambda_k\,\underset{\sim}{x}_k \quad \text{— (2)}$$

The transpose of eqn (1), and the complex conjugate of eqn (2), gives

$(\underset{\approx}{A}\,\underset{\sim}{x})^T = \underset{\sim}{x}^T\underset{\approx}{A}^T$

$$(1)^T \;\Rightarrow\; \underset{\sim}{x}_j^T\,\underset{\approx}{A}^T = \lambda_j\,\underset{\sim}{x}_j^T \quad \text{— (3)}$$

$\lambda^T = \lambda$

$$(2)^* \;\Rightarrow\; \underset{\approx}{A}^*\,\underset{\sim}{x}_k^* = \lambda_k^*\,\underset{\sim}{x}_k^* \quad \text{— (4)}$$

Then, subtracting eqn (4) pre-multiplied by $\underset{\sim}{x}_j^T$ from eqn (3) post-multiplied by $\underset{\sim}{x}_k^*$, we obtain

$$(3)\,\underset{\sim}{x}_k^* - \underset{\sim}{x}_j^T\,(4) \;\Rightarrow\; \underset{\sim}{x}_j^T\,\underset{\approx}{A}^T\,\underset{\sim}{x}_k^* - \underset{\sim}{x}_j^T\,\underset{\approx}{A}^*\,\underset{\sim}{x}_k^* = \lambda_j\,\underset{\sim}{x}_j^T\,\underset{\sim}{x}_k^* - \lambda_k^*\,\underset{\sim}{x}_j^T\,\underset{\sim}{x}_k^*$$

$$\text{But } \underset{\approx}{A}^T = \underset{\approx}{A}^* \;\Rightarrow\; (\lambda_j - \lambda_k^*)\,\underset{\sim}{x}_j^T\,\underset{\sim}{x}_k^* = 0$$

$$\text{If } j = k, \qquad \lambda_j = \lambda_j^* \qquad (\because\; \underset{\sim}{x}_j^T\,\underset{\sim}{x}_j^* > 0)$$

i.e. <u>The eigenvalues of an Hermitian matrix are real</u>

$\underset{\sim}{a}^T\underset{\sim}{b}^*$ is the general definition of the dot product between two potentially complex vectors $\underset{\sim}{a}$ and $\underset{\sim}{b}$; thus if $\underset{\sim}{a} = \underset{\sim}{b}$, then $\underset{\sim}{a}^T\underset{\sim}{a}^* \geq 0$ because it represents the square of the modulus (or magnitude) of $\underset{\sim}{a}$.

$$\text{If } j \neq k \text{ and } \lambda_j \neq \lambda_k, \qquad \underset{\sim}{x}_j^T\,\underset{\sim}{x}_k^* = 0$$

i.e. <u>The eigenvectors of distinct eigenvalues are orthogonal</u>

The condition $\underset{\sim}{a}^T\underset{\sim}{b}^* = 0$ for $\underset{\sim}{a} \neq \underset{\sim}{b}$ (with $\underset{\sim}{a} \neq 0$ and $\underset{\sim}{b} \neq 0$) is the definition of orthogonality, so that the vectors $\underset{\sim}{a}$ and $\underset{\sim}{b}$ are perpendicular to each other.

If $\lambda_j = \lambda_k$, so that two eigenvalues are the same, then it is known as a *degenerate* case. The related eigenvectors do not then specify a unique direction, but correspond to a pair of vectors that lie in a given plane. Since any point in a plane can be attained through a linear combination of two (non-collinear) basis vectors lying in it, we are at liberty to choose two orthogonal directions in the degeneracy plane as being suitable eigenvectors. The preceding analysis applies equally well to real and symmetric matrices, satisfying $\underset{\approx}{A} = \underset{\approx}{A}^*$ and $\underset{\approx}{A}^T = \underset{\approx}{A}$, since these are just a special case of Hermitian ones.

9.5 Find the eigenvalues and eigenvectors of

$$\underset{\approx}{A} = \begin{pmatrix} 1 & 0 & 1 \\ 0 & -1 & 0 \\ 1 & 0 & 1 \end{pmatrix}$$

Confirm that the eigenvectors are mutually orthogonal, the sum of the eigenvalues is equal to the trace of $\underset{\approx}{A}$, and the product of eigenvalues is equal to $\det(\underset{\approx}{A})$. Construct the diagonalisation matrix $\underset{\approx}{Q}$ from the normalised eigenvectors of $\underset{\approx}{A}$, and verify that $\underset{\approx}{Q}\,\underset{\approx}{Q}^T = \underset{\approx}{Q}^T\underset{\approx}{Q} = \underset{\approx}{I}$. Check that the *similarity* transform $\underset{\approx}{Q}^T\underset{\approx}{A}\,\underset{\approx}{Q} = \underset{\approx}{\Lambda}$ yields a diagonal matrix simply related to the eigenvalues of $\underset{\approx}{A}$.

Eigenvalue equation: $\underset{\approx}{A}\,\underset{\sim}{x} = \lambda\,\underset{\sim}{x} \quad \Rightarrow \quad (\underset{\approx}{A} - \lambda\,\underset{\approx}{I})\,\underset{\sim}{x} = 0$

$$x + z = \lambda x$$
$$-y = \lambda y$$
$$x + z = \lambda z$$

For non-trivial solutions, $\det(\underset{\approx}{A} - \lambda\,\underset{\approx}{I}) = 0$

$$\therefore \quad \begin{vmatrix} 1-\lambda & 0 & 1 \\ 0 & -1-\lambda & 0 \\ 1 & 0 & 1-\lambda \end{vmatrix} = 0$$

$$\therefore \quad (-1-\lambda)\begin{vmatrix} 1-\lambda & 1 \\ 1 & 1-\lambda \end{vmatrix} = 0$$

$$\therefore \quad (1+\lambda)(\lambda^2 - 2\lambda + 1 - 1) = 0$$

$$\therefore \quad \lambda(1+\lambda)(\lambda-2) = 0$$

i.e. Eigenvalues are: $\lambda = 0,\ \lambda = -1$ or $\lambda = 2$

$$\lambda_1 = 0$$
$$\lambda_2 = -1$$
$$\lambda_3 = 2$$

When $\lambda = 0$, $\left.\begin{matrix} x + z = 0 \\ y = 0 \end{matrix}\right\} \quad \begin{pmatrix} x \\ y \\ z \end{pmatrix} = \begin{pmatrix} t \\ 0 \\ -t \end{pmatrix}$

$\therefore$ Normalised eigenvector $\underset{\sim}{x}_1 = \frac{1}{\sqrt{2}}\begin{pmatrix} 1 \\ 0 \\ -1 \end{pmatrix}$

$$\text{When } \lambda = -1, \quad \left.\begin{aligned} 2x+z &= 0 \\ x+2z &= 0 \end{aligned}\right\} \quad \begin{pmatrix} x \\ y \\ z \end{pmatrix} = \begin{pmatrix} 0 \\ t \\ 0 \end{pmatrix}$$

$$\therefore \quad \underline{\text{Normalised eigenvector } \underset{\approx}{x}_2 = \begin{pmatrix} 0 \\ 1 \\ 0 \end{pmatrix}}$$

$$\text{When } \lambda = 2, \quad \left.\begin{aligned} -x+z &= 0 \\ -3y \; &= 0 \end{aligned}\right\} \quad \begin{pmatrix} x \\ y \\ z \end{pmatrix} = \begin{pmatrix} t \\ 0 \\ t \end{pmatrix}$$

$$\therefore \quad \underline{\text{Normalised eigenvector } \underset{\approx}{x}_3 = \frac{1}{\sqrt{2}} \begin{pmatrix} 1 \\ 0 \\ 1 \end{pmatrix}}$$

$$\underset{\approx}{x}_1 \cdot \underset{\approx}{x}_2 = \underset{\approx}{x}_1^T \underset{\approx}{x}_2 = \frac{1}{\sqrt{2}}[1 \times 0 + 0 \times 1 + (-1) \times 0] = 0$$

$$\text{Similarly,} \quad \underset{\approx}{x}_1^T \underset{\approx}{x}_3 = 0 \quad \text{and} \quad \underset{\approx}{x}_2^T \underset{\approx}{x}_3 = 0$$

$$\therefore \quad \underline{\text{Eigenvectors are mutually orthogonal}}$$

$$\text{trace}(\underset{\approx}{A}) = \text{sum of diagonal elements} = 1 - 1 + 1 = 1$$

$$\text{But, sum of eigenvalues} = 0 - 1 + 2 = 1$$

$$\therefore \quad \underline{\text{Sum of eigenvalues} = \text{trace}(\underset{\approx}{A})}$$

$$\det(\underset{\approx}{A}) = \begin{vmatrix} 1 & 0 & 1 \\ 0 & -1 & 0 \\ 1 & 0 & 1 \end{vmatrix} = -1 \times (1 - 1) = 0$$

$$\text{And, product of eigenvalues} = 0 \times (-1) \times 2 = 0$$

$$\therefore \quad \underline{\text{Product of eigenvalues} = \det(\underset{\approx}{A})}$$

$$\text{Diagonalisation matrix } \underset{\approx}{Q} = \begin{pmatrix} \frac{1}{\sqrt{2}} & 0 & \frac{1}{\sqrt{2}} \\ 0 & 1 & 0 \\ \frac{-1}{\sqrt{2}} & 0 & \frac{1}{\sqrt{2}} \end{pmatrix} = \frac{1}{\sqrt{2}} \begin{pmatrix} 1 & 0 & 1 \\ 0 & \sqrt{2} & 0 \\ -1 & 0 & 1 \end{pmatrix}$$

$$\therefore \quad \underset{\approx}{Q}\,\underset{\approx}{Q}^T = \frac{1}{2} \begin{pmatrix} 1 & 0 & 1 \\ 0 & \sqrt{2} & 0 \\ -1 & 0 & 1 \end{pmatrix} \begin{pmatrix} 1 & 0 & -1 \\ 0 & \sqrt{2} & 0 \\ 1 & 0 & 1 \end{pmatrix} = \frac{1}{2} \begin{pmatrix} 2 & 0 & 0 \\ 0 & 2 & 0 \\ 0 & 0 & 2 \end{pmatrix} = \underset{\approx}{I}$$

$$\underset{\approx}{Q}^T \underset{\approx}{Q} = \frac{1}{2} \begin{pmatrix} 1 & 0 & -1 \\ 0 & \sqrt{2} & 0 \\ 1 & 0 & 1 \end{pmatrix} \begin{pmatrix} 1 & 0 & 1 \\ 0 & \sqrt{2} & 0 \\ -1 & 0 & 1 \end{pmatrix} = \frac{1}{2} \begin{pmatrix} 2 & 0 & 0 \\ 0 & 2 & 0 \\ 0 & 0 & 2 \end{pmatrix} = \underset{\approx}{I}$$

$$\therefore \quad \underset{\approx}{Q} \text{ is orthogonal, because } \underline{\underset{\approx}{Q}\,\underset{\approx}{Q}^T = \underset{\approx}{Q}^T \underset{\approx}{Q} = \underset{\approx}{I}}$$

$$\underset{\approx}{Q}^T \underset{\approx}{A} = \frac{1}{\sqrt{2}} \begin{pmatrix} 1 & 0 & -1 \\ 0 & \sqrt{2} & 0 \\ 1 & 0 & 1 \end{pmatrix} \begin{pmatrix} 1 & 0 & 1 \\ 0 & -1 & 0 \\ 1 & 0 & 1 \end{pmatrix} = \frac{1}{\sqrt{2}} \begin{pmatrix} 0 & 0 & 0 \\ 0 & -\sqrt{2} & 0 \\ 2 & 0 & 2 \end{pmatrix}$$

$$\therefore \quad \underset{\approx}{Q}^T \underset{\approx}{A}\,\underset{\approx}{Q} = \frac{1}{2} \begin{pmatrix} 0 & 0 & 0 \\ 0 & -\sqrt{2} & 0 \\ 2 & 0 & 2 \end{pmatrix} \begin{pmatrix} 1 & 0 & 1 \\ 0 & \sqrt{2} & 0 \\ -1 & 0 & 1 \end{pmatrix} = \frac{1}{2} \begin{pmatrix} 0 & 0 & 0 \\ 0 & -2 & 0 \\ 0 & 0 & 4 \end{pmatrix}$$

$$\text{i.e. } \underset{\approx}{Q}^T \underset{\approx}{A}\,\underset{\approx}{Q} = \underline{\begin{pmatrix} 0 & 0 & 0 \\ 0 & -1 & 0 \\ 0 & 0 & 2 \end{pmatrix}} = \begin{pmatrix} \lambda_1 & 0 & 0 \\ 0 & \lambda_2 & 0 \\ 0 & 0 & \lambda_3 \end{pmatrix}$$

When evaluating the eigenvalues and eigenvectors of a real and symmetric matrix (or even an Hermitian one) it is always worth quickly checking that: **(i)** the sum of the eigenvalues is equal to the trace of the matrix; **(ii)** the eigenvectors are mutually orthogonal; and **(iii)** the product of the eigenvalues is equal to the determinant of the matrix (if this is not too difficult to calculate). A failure to meet any of these criteria is indicative of a mistake in the evaluation of the eigenproperties. Following the previous example, we should also make sure that all the eigenvalues are real.

10 Partial differentiation

10.1 Determine from first principles the partial derivatives $(\partial z/\partial x)_y$ and $(\partial z/\partial y)_x$ where $z(x,y) = x^3/(1-y)$. Evaluate (by any means) $\partial^2 z/\partial x^2$ and $\partial^2 z/\partial y^2$, and check that $\partial^2 z/\partial x\,\partial y = \partial^2 z/\partial y\,\partial x$.

$$\left(\frac{\partial z}{\partial x}\right)_y = \lim_{\delta x \to 0}\left(\frac{z(x+\delta x, y) - z(x,y)}{\delta x}\right)$$

$$= \lim_{\delta x \to 0}\left(\frac{(x+\delta x)^3/(1-y) - x^3/(1-y)}{\delta x}\right)$$

$$= \lim_{\delta x \to 0}\left(\frac{x^3 + 3x^2\,\delta x + 3x\,\delta x^2 + \delta x^3 - x^3}{(1-y)\,\delta x}\right)$$

$$= \lim_{\delta x \to 0}\left(\frac{3x^2 + 3x\,\delta x + \delta x^2}{(1-y)}\right) = \underline{\frac{3x^2}{1-y}}$$

$$\left(\frac{\partial z}{\partial y}\right)_x = \lim_{\delta y \to 0}\left(\frac{z(x, y+\delta y) - z(x,y)}{\delta y}\right)$$

$$= \lim_{\delta y \to 0}\left(\frac{x^3/(1-[y+\delta y]) - x^3/(1-y)}{\delta y}\right)$$

$$= \lim_{\delta y \to 0}\left(\frac{x^3}{\delta y}\left[\frac{1}{(1-y-\delta y)} - \frac{1}{(1-y)}\right]\right)$$

$$= \lim_{\delta y \to 0}\left(\frac{x^3\left[(1-y)-(1-y-\delta y)\right]}{\delta y\,(1-y-\delta y)(1-y)}\right)$$

$$= \lim_{\delta y \to 0}\left(\frac{x^3}{(1-y-\delta y)(1-y)}\right) = \underline{\frac{x^3}{(1-y)^2}}$$

$$\frac{\partial^2 z}{\partial x^2} = \frac{\partial}{\partial x_y}\left(\frac{\partial z}{\partial x}\right)_y = \frac{\partial}{\partial x_y}\left(\frac{3x^2}{1-y}\right) = \underline{\frac{6x}{1-y}}$$

$$\frac{\partial^2 z}{\partial y^2} = \frac{\partial}{\partial y_x}\left(\frac{\partial z}{\partial y}\right)_x = \frac{\partial}{\partial y_x}\left(\frac{x^3}{(1-y)^2}\right) = \underline{\frac{2x^3}{(1-y)^3}}$$

$$\frac{\partial^2 z}{\partial x\,\partial y} = \frac{\partial}{\partial x_y}\left(\frac{\partial z}{\partial y}\right)_x = \frac{\partial}{\partial x_y}\left(\frac{x^3}{(1-y)^2}\right) = \underline{\frac{3x^2}{(1-y)^2}}$$

$$\frac{\partial^2 z}{\partial y\,\partial x} = \frac{\partial}{\partial y_x}\left(\frac{\partial z}{\partial x}\right)_y = \frac{\partial}{\partial y_x}\left(\frac{3x^2}{1-y}\right) = \underline{\frac{3x^2}{(1-y)^2}}$$

Although we know that the mixed second derivatives should be equal, it's usually worth making sure that they are as a simple check on our calculation.

10.2 If $f(x,y,z) = \cos(xyz)$, evaluate $\partial^3 f/\partial x\partial y\partial z$ in which the appropriate variables are held constant.

$$\frac{\partial f}{\partial z} = \left(\frac{\partial f}{\partial z}\right)_{xy} = -xy\sin(xyz)$$

$$\therefore \quad \frac{\partial^2 f}{\partial y\,\partial z} = \frac{\partial}{\partial y_{xz}}\left(\frac{\partial f}{\partial z}\right)_{xy} = -x^2yz\cos(xyz) - x\sin(xyz)$$

$$\therefore \quad \frac{\partial^3 f}{\partial x\,\partial y\,\partial z} = \frac{\partial}{\partial x_{yz}}\left(\frac{\partial^2 f}{\partial y\,\partial z}\right) = x^2y^2z^2\sin(xyz) - 2xyz\cos(xyz)$$

$$-xyz\cos(xyz) - \sin(xyz)$$

$$= \underline{(x^2y^2z^2 - 1)\sin(xyz) - 3xyz\cos(xyz)}$$

10.3 Verify that $x^2 = y^2\sin(yz)$ satisfies $(\partial x/\partial y)_z(\partial y/\partial z)_x(\partial z/\partial x)_y = -1$.

$$x^2 = y^2\sin(yz) \quad \text{— (1)}$$

$$\frac{\partial}{\partial y_z}(1) \Rightarrow \quad 2x\left(\frac{\partial x}{\partial y}\right)_z = y^2\cos(yz)\frac{\partial}{\partial y_z}(yz) + 2y\sin(yz)$$

$$\therefore \quad \left(\frac{\partial x}{\partial y}\right)_z = \frac{y^2z\cos(yz) + 2y\sin(yz)}{2x} \quad \text{— (2)}$$

$$\frac{\partial}{\partial z_x}(1) \Rightarrow \quad 0 = y^2\cos(yz)\frac{\partial}{\partial z_x}(yz) + 2y\left(\frac{\partial y}{\partial z}\right)_x\sin(yz)$$

$$= y^2\cos(yz)\left[y + z\left(\frac{\partial y}{\partial z}\right)_x\right] + 2y\left(\frac{\partial y}{\partial z}\right)_x\sin(yz)$$

$$\therefore \quad \left(\frac{\partial y}{\partial z}\right)_x = \frac{-y^3\cos(yz)}{y^2z\cos(yz) + 2y\sin(yz)} \quad \text{— (3)}$$

$$\frac{\partial}{\partial x_y}(1) \Rightarrow \qquad 2x = y^2\cos(yz)\frac{\partial}{\partial x_y}(yz) = y^3\cos(yz)\left(\frac{\partial z}{\partial x}\right)_y$$

$$\therefore \quad \left(\frac{\partial z}{\partial x}\right)_y = \frac{2x}{y^3\cos(yz)} \qquad \text{— (4)}$$

$$\therefore \quad (2)\times(3)\times(4) \Rightarrow \quad \underline{\left(\frac{\partial x}{\partial y}\right)_z\left(\frac{\partial y}{\partial z}\right)_x\left(\frac{\partial z}{\partial x}\right)_y = -1}$$

10.4 Consider the function $f(x,y) = xy(1-y+x)$. Calculate the gradient vector, $\underset{\sim}{\nabla}f$, at the points $(-1/2,0)$, $(-1/2,1/2)$ and $(0,1/2)$. Find the stationary point which lies within the triangle bounded by the points $(-1,0)$, $(0,0)$ and $(0,1)$. Sketch the function within this triangle, and mark in the directions of $\underset{\sim}{\nabla}f$ where it has been calculated.

$$\underset{\sim}{\nabla}f = \left(\frac{\partial f}{\partial x}, \frac{\partial f}{\partial y}\right), \quad \text{where} \quad \left(\frac{\partial f}{\partial x}\right)_y = \frac{\partial}{\partial x_y}(xy - xy^2 + x^2y) = y(1-y+2x)$$

$$\text{and} \quad \left(\frac{\partial f}{\partial y}\right)_x = \frac{\partial}{\partial y_x}(xy - xy^2 + x^2y) = x(1-2y+x)$$

$$\therefore \quad \underset{\sim}{\nabla}f\left(-\tfrac{1}{2},0\right) = \underline{\left(0,-\tfrac{1}{4}\right)}; \quad \underset{\sim}{\nabla}f\left(-\tfrac{1}{2},\tfrac{1}{2}\right) = \underline{\left(-\tfrac{1}{4},\tfrac{1}{4}\right)}; \quad \underset{\sim}{\nabla}f\left(0,\tfrac{1}{2}\right) = \underline{\left(\tfrac{1}{4},0\right)}$$

For stationary point, $\underset{\sim}{\nabla}f = 0$

$$\left.\begin{aligned} \therefore \quad & y(1-y+2x) = 0 \\ \text{and} \quad & x(1-2y+x) = 0 \end{aligned}\right\} \quad \begin{aligned} & -1-3x = 0 \\ \therefore \quad & x = -1/3 \ \text{ and } \ y = 1/3 \end{aligned}$$

$\therefore$ Stationary point inside triangle is at $\underline{\left(-\tfrac{1}{3},\tfrac{1}{3}\right)}$.

The nature of the stationary point can be ascertained as being a minimum without examining the second derivatives: for example, the gradient vectors point outwards everywhere; or, the value of f at $(-1/3,1/3)$ is $-1/27$ while the edge of the triangle is a contour with $f=0$.

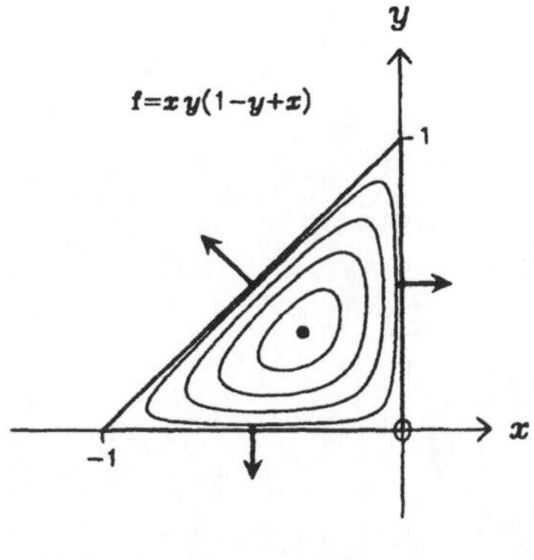

10.5 Show that if $f(u,v)=0$, where $u=x+y$ and $v=x^2+xy+z^2$, then $x+y=2z[(\partial z/\partial y)_x-(\partial z/\partial x)_y]$.

$$f = f(u,v) = 0 \quad \Rightarrow \quad df = \left(\frac{\partial f}{\partial u}\right)_v du + \left(\frac{\partial f}{\partial v}\right)_u dv = 0$$

$$\therefore \quad \left(\frac{\partial f}{\partial u}\right)_v \left(\frac{\partial u}{\partial x}\right)_y = -\left(\frac{\partial f}{\partial v}\right)_u \left(\frac{\partial v}{\partial x}\right)_y$$

$$\text{i.e.} \quad \left(\frac{\partial f}{\partial u}\right)_v = -\left(\frac{\partial f}{\partial v}\right)_u \left[2x+y+2z\left(\frac{\partial z}{\partial x}\right)_y\right] \quad \text{— (1)}$$

$$\text{Also,} \quad \left(\frac{\partial f}{\partial u}\right)_v \left(\frac{\partial u}{\partial y}\right)_x = -\left(\frac{\partial f}{\partial v}\right)_u \left(\frac{\partial v}{\partial y}\right)_x$$

$$\text{i.e.} \quad \left(\frac{\partial f}{\partial u}\right)_v = -\left(\frac{\partial f}{\partial v}\right)_u \left[x+2z\left(\frac{\partial z}{\partial y}\right)_x\right] \quad \text{— (2)}$$

$$(1) \div (2) \quad \Rightarrow \quad 2x+y+2z\left(\frac{\partial z}{\partial x}\right)_y = x+2z\left(\frac{\partial z}{\partial y}\right)_x$$

$$\text{i.e.} \quad \underline{x+y = 2z\left[\left(\frac{\partial z}{\partial y}\right)_x - \left(\frac{\partial z}{\partial x}\right)_y\right]}$$

10.6 Use the substitution $u=x+ct$ and $v=x-ct$ to reduce the wave equation $c^2\partial^2 z/\partial x^2 = \partial^2 z/\partial t^2$ to the form $\partial^2 z/\partial u\,\partial v = 0$.

$$z = z(u,v) \quad \Rightarrow \quad dz = \left(\frac{\partial z}{\partial u}\right)_v du + \left(\frac{\partial z}{\partial v}\right)_u dv$$

$$\therefore \quad \left(\frac{\partial z}{\partial x}\right)_t = \left(\frac{\partial z}{\partial u}\right)_v \left(\frac{\partial u}{\partial x}\right)_t + \left(\frac{\partial z}{\partial v}\right)_u \left(\frac{\partial v}{\partial x}\right)_t = \left(\frac{\partial z}{\partial u}\right)_v + \left(\frac{\partial z}{\partial v}\right)_u$$

$$\text{and} \quad \left(\frac{\partial z}{\partial t}\right)_x = \left(\frac{\partial z}{\partial u}\right)_v \left(\frac{\partial u}{\partial t}\right)_x + \left(\frac{\partial z}{\partial v}\right)_u \left(\frac{\partial v}{\partial t}\right)_x = c\left(\frac{\partial z}{\partial u}\right)_v - c\left(\frac{\partial z}{\partial v}\right)_u$$

$$\frac{\partial^2 z}{\partial x^2} = \frac{\partial}{\partial x_t}\left(\frac{\partial z}{\partial x}\right)_t = \left[\frac{\partial}{\partial u_v} + \frac{\partial}{\partial v_u}\right]\left[\left(\frac{\partial z}{\partial u}\right)_v + \left(\frac{\partial z}{\partial v}\right)_u\right]$$

$$= \frac{\partial}{\partial u_v}\left(\frac{\partial z}{\partial u}\right)_v + \frac{\partial}{\partial u_v}\left(\frac{\partial z}{\partial v}\right)_u + \frac{\partial}{\partial v_u}\left(\frac{\partial z}{\partial u}\right)_v + \frac{\partial}{\partial v_u}\left(\frac{\partial z}{\partial v}\right)_u$$

$$\therefore \quad \frac{\partial^2 z}{\partial x^2} = \frac{\partial^2 z}{\partial u^2} + 2\frac{\partial^2 z}{\partial u\,\partial v} + \frac{\partial^2 z}{\partial v^2} \quad \text{— (1)}$$

$$\frac{\partial^2 z}{\partial t^2} = \frac{\partial}{\partial t_x}\left(\frac{\partial z}{\partial t}\right)_x = \left[c\frac{\partial}{\partial u_v} - c\frac{\partial}{\partial v_u}\right]\left[c\left(\frac{\partial z}{\partial u}\right)_v - c\left(\frac{\partial z}{\partial v}\right)_u\right]$$

$$= c\frac{\partial}{\partial u_v}\left[c\left(\frac{\partial z}{\partial u}\right)_v - c\left(\frac{\partial z}{\partial v}\right)_u\right] - c\frac{\partial}{\partial v_u}\left[c\left(\frac{\partial z}{\partial u}\right)_v - c\left(\frac{\partial z}{\partial v}\right)_u\right]$$

$$\therefore \quad \frac{\partial^2 z}{\partial t^2} = c^2\frac{\partial^2 z}{\partial u^2} - 2c^2\frac{\partial^2 z}{\partial u\,\partial v} + c^2\frac{\partial^2 z}{\partial v^2} \quad \text{— (2)}$$

$$(1) - \frac{(2)}{c^2} \quad \Rightarrow \quad \frac{\partial^2 z}{\partial x^2} - \frac{1}{c^2}\frac{\partial^2 z}{\partial t^2} = 4\frac{\partial^2 z}{\partial u\,\partial v}$$

$$\text{But} \quad \frac{\partial^2 z}{\partial x^2} - \frac{1}{c^2}\frac{\partial^2 z}{\partial t^2} = 0 \qquad \therefore \quad \underline{\frac{\partial^2 z}{\partial u\,\partial v} = 0}$$

The partial differential operators, $\partial/\partial x_t$ and $\partial/\partial t_x$, are easily ascertained from the expressions for the first derivatives, $(\partial z/\partial x)_t$ and $(\partial z/\partial t)_x$, as long as we are careful to rearrange the equations so that z always appears on the far right. For example,

$$\left(\frac{\partial z}{\partial t}\right)_x = \frac{\partial}{\partial t_x}(z) = c\frac{\partial}{\partial u_v}(z) - c\frac{\partial}{\partial v_u}(z) = \left[c\frac{\partial}{\partial u_v} - c\frac{\partial}{\partial v_u}\right](z)$$

$$\therefore \quad \frac{\partial}{\partial t_x} = c\frac{\partial}{\partial u_v} - c\frac{\partial}{\partial v_u}$$

Differential operators obey the same rules as ordinary algebraic expressions, such as for multiplying out brackets and so on, and apply to everything on their immediate right. The wave equation case was easy because c was a constant; if its equivalent had been a function of x and y, and therefore (implicitly) one of u and v, then we would have needed to use the product rule several times, as in

$$\phi(u,v)\frac{\partial}{\partial u_v}\left[\phi(u,v)\left(\frac{\partial z}{\partial u}\right)_v\right] = \phi(u,v)\left[\phi(u,v)\frac{\partial^2 z}{\partial u^2} + \left(\frac{\partial \phi}{\partial u}\right)_v\left(\frac{\partial z}{\partial u}\right)_v\right]$$

so that the related algebra (but not the principle of the procedure) can become rather messy.

10.7 Find and classify all the (real) stationary values of the function $f(x,y) = y^2(a^2+x^2) - x^2(2a^2-x^2)$, where a is a constant.

$$\left(\frac{\partial f}{\partial x}\right)_y = 2xy^2 - 4a^2x + 4x^3 = 2x(y^2 - 2a^2 + 2x^2)$$

$$\left(\frac{\partial f}{\partial y}\right)_x = 2y(a^2+x^2)$$

$$\therefore \quad \frac{\partial^2 f}{\partial x^2} = 2y^2 - 4a^2 + 12x^2\,; \quad \frac{\partial^2 f}{\partial y^2} = 2(a^2+x^2)\,; \quad \frac{\partial^2 f}{\partial x\,\partial y} = 4xy$$

For stationary points, $\left(\frac{\partial f}{\partial x}\right)_y = 0$ and $\left(\frac{\partial f}{\partial y}\right)_x = 0$

$$\therefore \quad \left.\begin{aligned} x(y^2-2a^2+2x^2) &= 0 \\ \text{and} \quad y(a^2+x^2) &= 0 \end{aligned}\right\} \quad \begin{aligned} &x=0 \text{ and } y=0 \\ \text{or } &y=0 \text{ and } x=\pm a \end{aligned}$$

To classify stationary points, first need to consider sign of $\det(\underset{\sim}{\nabla}\underset{\sim}{\nabla} f)$

$$\text{where} \quad \det(\underset{\sim}{\nabla}\underset{\sim}{\nabla} f) = \left(\frac{\partial^2 f}{\partial x^2}\right)\left(\frac{\partial^2 f}{\partial y^2}\right) - \left(\frac{\partial^2 f}{\partial x \partial y}\right)^2$$

At $(0,0)$, $\det(\underset{\sim}{\nabla}\underset{\sim}{\nabla} f) = -8a^4 < 0$ $\therefore$ <u>$(0,0)$ is a saddle point</u>

$[\text{real } a \Rightarrow a^4 \geq 0]$

At $(\pm a,0)$, $\det(\underset{\sim}{\nabla}\underset{\sim}{\nabla} f) = 32a^2 > 0$ $\therefore$ maximum or minimum

i.e. Need sign of $\partial^2 f/\partial x^2$, or $\partial^2 f/\partial y^2$, or $\nabla^2 f = \partial^2 f/\partial x^2 + \partial^2 f/\partial y^2$

At $(+a,0)$, $\nabla^2 f = 12a^2 > 0$ $\therefore$ <u>$(a,0)$ is a minimum</u>

At $(-a,0)$, $\nabla^2 f = 12a^2 > 0$ $\therefore$ <u>$(-a,0)$ is a minimum</u>

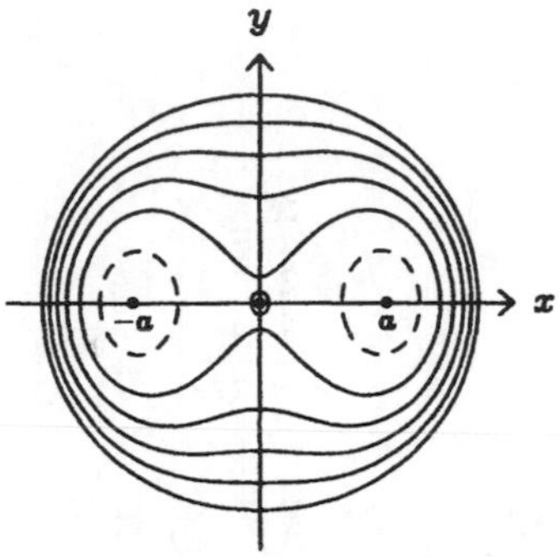

When trying to find the stationary points, it is usually well worth factorising the first derivatives as much as possible first. In this example, $(\partial f/\partial x)_y = 0$ tells us that either $x = 0$ or $y^2 - 2a^2 + 2x^2 = 0$ and $(\partial f/\partial y)_x = 0$ means that either $y = 0$ or $a^2 + x^2 = 0$. By considering each of the four combinations which make both the first derivatives simultaneously equal to zero in turn,we are assured of finding all the stationary points.

10.8 Using the method of Lagrange multipliers, find the stationary values of the function e^{-xy} subject to the condition $x^2 + y^2 = 1$.

For stationary points of $f(x,y) = e^{-xy}$ subject to $g(x,y) = x^2 + y^2 - 1 = 0$

$F = e^{-xy} + \lambda(x^2 + y^2 - 1)$

Need $\left(\frac{\partial F}{\partial x}\right)_y = 0$ and $\left(\frac{\partial F}{\partial y}\right)_x = 0$, where $F(x,y) = f(x,y) + \lambda g(x,y)$

But $\left(\frac{\partial F}{\partial x}\right)_y = -y\,e^{-xy} + 2x\lambda$ and $\left(\frac{\partial F}{\partial y}\right)_x = -x\,e^{-xy} + 2y\lambda$

$$\therefore \quad y\,e^{-xy} = 2x\lambda \quad \text{— (1)}$$

$$\text{and} \quad x\,e^{-xy} = 2y\lambda \quad \text{— (2)}$$

$$g = 0 \Rightarrow \quad x^2 + y^2 = 1 \quad \text{— (3)}$$

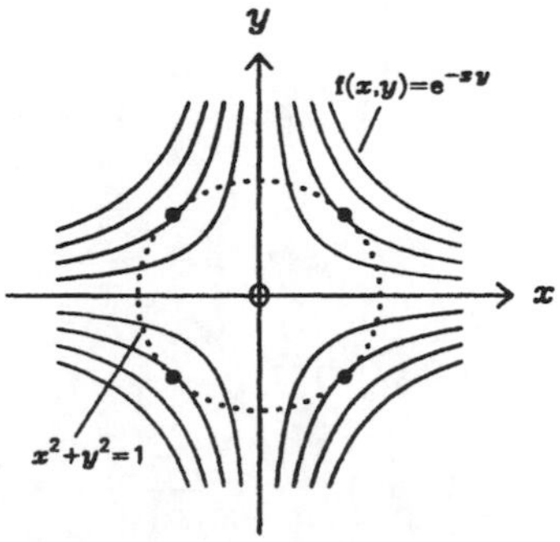

$$(1) \div (2) \;\Rightarrow\; \frac{y}{x} = \frac{x}{y} \qquad \therefore\; x^2 - y^2 = 0 \qquad \text{— (4)}$$

$$(3) + (4) \;\Rightarrow\; 2x^2 = 1 \qquad \therefore\; x = \pm 1/\sqrt{2}\,, \text{ and } (4) \;\Rightarrow\; y = \pm x$$

$$\therefore\; \text{Stationary points are at } \underline{\pm\left(1/\sqrt{2}, 1/\sqrt{2}\right)}\,, \quad \text{with } \mathrm{f} = \mathrm{e}^{-1/2}$$

$$\text{and at } \underline{\pm\left(1/\sqrt{2}, -1/\sqrt{2}\right)}\,, \quad \text{with } \mathrm{f} = \mathrm{e}^{1/2}$$

> **10.9** By using the multivariate form of the Taylor series, derive a version of the Newton-Raphson algorithm for numerically finding a stationary point of a multiparameter function $\mathrm{f}(\underset{\sim}{x})$.

If $\mathrm{f} = \mathrm{f}(x,y)$

$$\underset{\sim}{\nabla}\mathrm{f} = \begin{pmatrix} \dfrac{\partial \mathrm{f}}{\partial x} \\ \dfrac{\partial \mathrm{f}}{\partial y} \end{pmatrix}$$

$$\underset{\sim}{\nabla}\underset{\sim}{\nabla}\mathrm{f} = \begin{pmatrix} \dfrac{\partial^2 \mathrm{f}}{\partial x^2} & \dfrac{\partial^2 \mathrm{f}}{\partial x \partial y} \\ \dfrac{\partial^2 \mathrm{f}}{\partial y \partial x} & \dfrac{\partial^2 \mathrm{f}}{\partial y^2} \end{pmatrix}$$

Using 'matrix-vector' notation, the Taylor series can be generalised as

$$\mathrm{f}(\underset{\sim}{x}) = \mathrm{f}(\underset{\sim}{x}_0) + (\underset{\sim}{x} - \underset{\sim}{x}_0)^T \underset{\sim}{\nabla}\mathrm{f}(\underset{\sim}{x}_0) + \tfrac{1}{2}(\underset{\sim}{x} - \underset{\sim}{x}_0)^T \underset{\sim}{\nabla}\underset{\sim}{\nabla}\mathrm{f}(\underset{\sim}{x}_0)(\underset{\sim}{x} - \underset{\sim}{x}_0) + \cdots$$

For stationary points, $\underset{\sim}{\nabla}\mathrm{f}(\underset{\sim}{x}) = 0$. Therefore

$$\underset{\sim}{\nabla}\mathrm{f}(\underset{\sim}{x}) = \underset{\sim}{\nabla}\mathrm{f}(\underset{\sim}{x}_0) + \underset{\sim}{\nabla}\underset{\sim}{\nabla}\mathrm{f}(\underset{\sim}{x}_0)(\underset{\sim}{x} - \underset{\sim}{x}_0) + \cdots = 0$$

If this multivariate differentiation feels awkward, think of it as a Taylor series expansion of $\underset{\sim}{\nabla}\mathrm{f}$ (rather than f). If $\underset{\sim}{x}_0$ is a 'good' estimate of the solution of $\underset{\sim}{\nabla}\mathrm{f}(\underset{\sim}{x}) = 0$, so that higher-order terms are negligible, then

$$\underset{\sim}{\nabla}\mathrm{f}(\underset{\sim}{x}_0) + \underset{\sim}{\nabla}\underset{\sim}{\nabla}\mathrm{f}(\underset{\sim}{x}_0)(\underset{\sim}{x} - \underset{\sim}{x}_0) \approx 0$$

$$\therefore\; \left[\underset{\sim}{\nabla}\underset{\sim}{\nabla}\mathrm{f}(\underset{\sim}{x}_0)\right]^{-1}\left[\underset{\sim}{\nabla}\mathrm{f}(\underset{\sim}{x}_0) + \underset{\sim}{\nabla}\underset{\sim}{\nabla}\mathrm{f}(\underset{\sim}{x}_0)(\underset{\sim}{x} - \underset{\sim}{x}_0)\right] \approx 0$$

$$\therefore\; \left[\underset{\sim}{\nabla}\underset{\sim}{\nabla}\mathrm{f}(\underset{\sim}{x}_0)\right]^{-1}\underset{\sim}{\nabla}\mathrm{f}(\underset{\sim}{x}_0) + \underset{\approx}{I}(\underset{\sim}{x} - \underset{\sim}{x}_0) \approx 0$$

$$\therefore\; \left[\underset{\sim}{\nabla}\underset{\sim}{\nabla}\mathrm{f}(\underset{\sim}{x}_0)\right]^{-1}\underset{\sim}{\nabla}\mathrm{f}(\underset{\sim}{x}_0) \approx \underset{\sim}{x}_0 - \underset{\sim}{x} \quad\Rightarrow\quad \underset{\sim}{x} \approx \underset{\sim}{x}_0 - \left[\underset{\sim}{\nabla}\underset{\sim}{\nabla}\mathrm{f}(\underset{\sim}{x}_0)\right]^{-1}\underset{\sim}{\nabla}\mathrm{f}(\underset{\sim}{x}_0)$$

In other words, a better estimate of the solution of $\underset{\sim}{\nabla}\mathrm{f}(\underset{\sim}{x}) = 0$ can be obtained from the values of the gradient vector and the second-derivative matrix at $\underset{\sim}{x}_0$. This forms the basis of an iterative Newton-Raphson algorithm:

$$\underline{\underset{\sim}{x}_{N+1} = \underset{\sim}{x}_N - \left[\underset{\sim}{\nabla}\underset{\sim}{\nabla}\mathrm{f}(\underset{\sim}{x}_N)\right]^{-1}\underset{\sim}{\nabla}\mathrm{f}(\underset{\sim}{x}_N)}$$

If $\mathrm{f} = \mathrm{f}(x_1, x_2, x_3, \cdots, x_M)$

$$(\underset{\sim}{\nabla}\mathrm{f})_j = \frac{\partial \mathrm{f}}{\partial x_j}$$

$$(\underset{\sim}{\nabla}\underset{\sim}{\nabla}\mathrm{f})_{jk} = \frac{\partial^2 \mathrm{f}}{\partial x_j \partial x_k}$$

where $\underset{\sim}{x}_N$ is the Nth estimate, and $\underset{\sim}{x}_{N+1}$ is the next improvement upon it.

The Newton-Raphson algorithm above is usually by far the most efficient way of numerically homing in on a stationary point given a good initial guess. The last proviso is important, though, because the procedure can easily diverge rapidly away from the solution being sought if the initial estimate is not close-enough to the desired point.

11 Line integrals

11.1 Show that the integral of $y^3\,dx + 3xy^2\,dy$ is independent of path by evaluating it over the following two routes between the origin and the point $(1,1)$: **(a)** $y = x^2$; and **(b)** the straight lines from $(0,0)$ to $(1,0)$, and from $(1,0)$ to $(1,1)$.

(a) Along $y = x^2$, $dy = 2x\,dx$

$$\therefore \quad \int_{\text{path(a)}} y^3\,dx + 3xy^2\,dy = \int_{x=0}^{x=1} (x^6 + 6x^6)\,dx = \Big[x^7\Big]_0^1 = \underline{1}$$

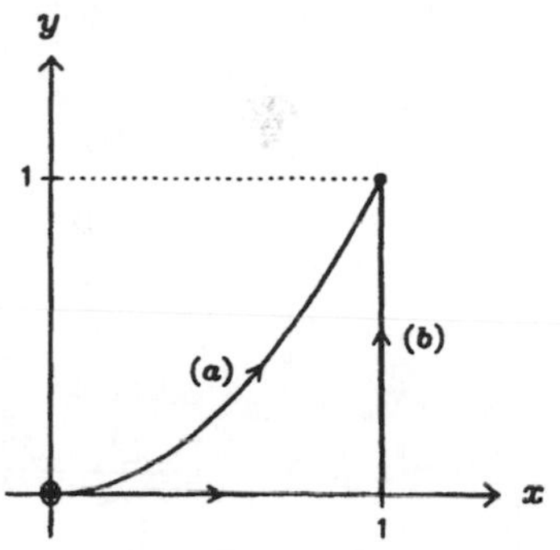

(b) From $(0,0)$ to $(1,0)$, $y = 0$ and $dy = 0$

$(1,0)$ to $(1,1)$, $x = 1$ and $dx = 0$

$$\therefore \quad \int_{\text{path(b)}} y^3\,dx + 3xy^2\,dy = 0 + \int_{y=0}^{y=1} 3y^2\,dy = \Big[y^3\Big]_0^1 = \underline{1}$$

Although the evaluation of the integral of $y^3\,dx + 3xy^2\,dy$ over the two paths, (a) and (b), is a useful verification that it is independent of path, the formal proof comes from the satisfaction of the condition that

$$\frac{\partial}{\partial y_x}\left(y^3\right) = \frac{\partial}{\partial x_y}\left(3xy^2\right)$$

11.2 Evaluate the integral of $xy\,dl$ along the two paths, (a) and (b), of the previous example.

$$dl = \sqrt{1 + \left(\frac{dy}{dx}\right)^2}\,dx \qquad \text{(by Pythagoras)}$$

(a) Along $y = x^2$, $\frac{dy}{dx} = 2x$

$$\therefore \quad \int_{\text{path(a)}} xy\,dl = \int_{x=0}^{x=1} x^3\sqrt{1+4x^2}\,dx$$

Put $u^2 = 1+4x^2 \qquad \therefore \; 2u\,du = 8x\,dx$

$$\therefore \quad \int_{\text{path(a)}} xy\,dl = \int_{u=1}^{u=\sqrt{5}} \frac{(u^2-1)}{4}\,u\,\frac{u\,du}{4}$$

$$= \frac{1}{16}\int_1^{\sqrt{5}} (u^4 - u^2)\,du$$

$$= \frac{1}{16}\left[\frac{u^5}{5} - \frac{u^3}{3}\right]_1^{\sqrt{5}}$$

$$= \frac{1}{16}\left[5\sqrt{5} - \frac{5\sqrt{5}}{3} - \frac{1}{5} + \frac{1}{3}\right]$$

$$= \frac{1}{16} \times \frac{(75-25)\sqrt{5} - 3 + 5}{15}$$

$$= \frac{50\sqrt{5}+2}{16\times 15} = \underline{\frac{25\sqrt{5}+1}{120}}$$

(b) From (0,0) to (1,0), $y = 0$ and $dl = dx$
(1,0) to (1,1), $x = 1$ and $dl = dy$

$$\therefore \quad \int_{\text{path(b)}} xy\,dl = 0 + \int_{y=0}^{y=1} y\,dy = \left[\frac{y^2}{2}\right]_0^1 = \underline{\frac{1}{2}}$$

11.3 Prove that if C_V is independent of volume V, $\delta q = C_V dT + (RT/V)dV$ is not an exact differential (R being a constant). Show that by dividing this equation by T, it becomes exact. Comment on the relevance of this to thermodynamics.

C_V independent of $V \;\Rightarrow\; \frac{\partial}{\partial V_T}\left(C_V\right) = 0$

$$\text{But} \quad \frac{\partial}{\partial T_V}\left(\frac{RT}{V}\right) = \frac{R}{V} \neq 0 \quad \Rightarrow \quad \frac{\partial}{\partial V_T}\left(C_V\right) \neq \frac{\partial}{\partial T_V}\left(\frac{RT}{V}\right)$$

$$\therefore \quad \underline{\delta q = C_V\,\mathrm{d}T + \frac{RT}{V}\,\mathrm{d}V \quad \text{is not exact}}$$

$$\frac{\partial}{\partial V_T}\left(\frac{C_V}{T}\right) = 0 \quad \text{and} \quad \frac{\partial}{\partial T_V}\left(\frac{R}{V}\right) = 0$$

$$\therefore \quad \underline{\frac{\delta q}{T} = \frac{C_V}{T}\,\mathrm{d}T + \frac{R}{V}\,\mathrm{d}V \quad \text{is exact}}$$

In thermodynamics, δq represents the change in heat. Since this is not an exact differential, the heat of the system does not only depend on the present and starting conditions but also on the details of how it go there; were the pressure and volume held fixed first and the temperature allowed to change, for example, or the other way around in going from (P_1, V_1, T_1) to (P_2, V_2, T_2). When divided by the temperature, $\delta q/T$ represents the change in entropy. As this is an exact differential, so that $\int \delta q/T$ is independent of the specific way in which the system was altered in going from (P_1, V_1, T_1) to (P_2, V_2, T_2), entropy is said to be a state function: its value depends on the state that it's in, and not on how it got there.

Incidentally, $1/T$ is the *integrating factor* for this problem: the multiplicative 'weighting term' which converts an inexact differential into an exact one. We could have derived it knowing only that it was a function of temperature, $w(T)$, from the condition that

$$\frac{\partial}{\partial V_T}\left[C_V\,w(T)\right] = \frac{\partial}{\partial T_V}\left[\frac{RT}{V}\,w(T)\right] \quad \Rightarrow \quad 0 = \frac{RT}{V}\frac{\mathrm{d}w}{\mathrm{d}T} + \frac{R}{V}\,w$$

where we have replaced $(\partial w/\partial T)_V$ by $\mathrm{d}w/\mathrm{d}T$ because, by definition, w depends only on T. This yields a simple first-order ordinary differential equation, which we will meet in Chapter 13, and has the solution $w = A/T$ where A is a constant.

12 Multiple integrals

12.1 Show that the area of the ellipse $(x/a)^2+(y/b)^2=1$ is equal to πab.

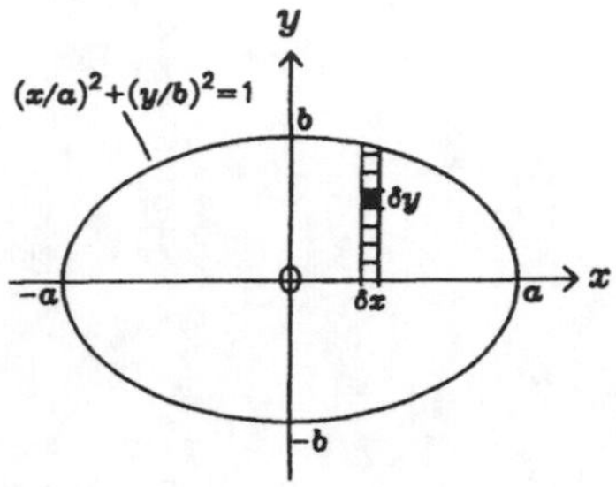

$$\text{Area} = \iint\limits_{\text{ellipse}} \mathrm{d}x\,\mathrm{d}y = 4\int_{x=0}^{x=a}\mathrm{d}x \int_{y=0}^{b\sqrt{1-(x/a)^2}}\mathrm{d}y = 4\int_0^a \Big[y\Big]_0^{b\sqrt{1-(x/a)^2}}\mathrm{d}x$$

$$= \frac{4b}{a}\int_0^a \sqrt{a^2-x^2}\,\mathrm{d}x$$

$[x = a\sin\theta]$

$$\text{But} \quad \int_0^a \sqrt{a^2-x^2}\,\mathrm{d}x = a^2\int_0^{\pi/2}\cos^2\theta\,\mathrm{d}\theta = \frac{a^2}{2}\int_0^{\pi/2}(\cos 2\theta+1)\,\mathrm{d}\theta = \frac{\pi a^2}{4}$$

$$\therefore \quad \underline{\text{Area} = \pi ab}$$

Early in the calculation, we replaced the integral over the ellipse by four times that of the positive quadrant; this kind of manipulation is often useful in problems with a lot of symmetry, as it helps to avoid subsequent difficulties arising from the limits of the integrals. It is also good practice to check derivations by making sure that they lead to sensible, or well-known, results under simplifying conditions. For example, an ellipse becomes a circle when $a=b$; our formula then correctly reduces to the familiar form πa^2, where a is the radius.

12.2 By drawing a suitable diagram, show that an element of volume in spherical polar coordinates is given by $r^2\sin\theta\,dr\,\mathrm{d}\theta\,\mathrm{d}\phi$; hence derive the formula for the volume of a sphere.

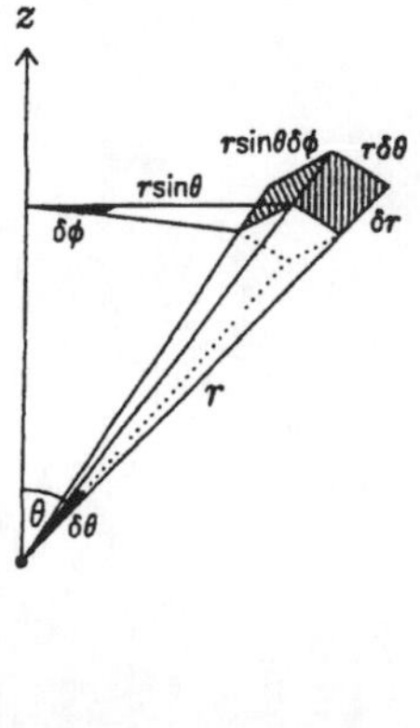

$$\text{Volume of small 'cuboid' element} = \delta r \times r\delta\theta \times r\sin\theta\,\delta\phi$$

$$\therefore \quad \text{Volume element dVol} = \underline{r^2\sin\theta\,dr\,\mathrm{d}\theta\,\mathrm{d}\phi}$$

$$\therefore \quad \text{Volume of sphere} = \iiint\limits_{sphere}\mathrm{dVol} = \int_{r=0}^{R} r^2\,\mathrm{d}r\int_{\theta=0}^{\theta=\pi}\sin\theta\,\mathrm{d}\theta\int_{\phi=0}^{\phi=2\pi}\mathrm{d}\phi$$

$$\therefore \text{ Volume } = \left[\frac{r^3}{3}\right]_0^R \times \left[-\cos\theta\right]_0^\pi \times \left[\phi\right]_0^{2\pi} = \frac{R^3}{3} \times (1+1) \times 2\pi$$

$$= \underline{\tfrac{4}{3}\pi R^3}$$

Although this triple integral is easy to evaluate, because spherical polar coordinates match the natural geometry of the object, and so the problem reduces to a simple product of three one-dimensional integrals, we will see an alternative way of doing the calculation in the next example which exploits the symmetry of the sphere from the outset.

Incidentally, our diagram also enables us to write down the expression for an element of the external area and, thereby, derive the formula for the surface area of a sphere:

$$\text{Surface area} = \iint_{sphere} \text{dArea} = R^2 \int_{\theta=0}^{\theta=\pi} \sin\theta \, d\theta \int_{\phi=0}^{\phi=2\pi} d\phi = \underline{4\pi R^2}$$

$$\text{dArea} = R^2 \sin\theta \, d\theta \, d\phi$$

> **12.3** By using cylindrical polar coordinates, show that the volume of the solid generated when the curve $y = f(x)$, for $a \le x \le b$, is rotated about the x-axis through 360° is given by $\pi \int_a^b y^2 \, dx$.

$$\text{Volume element in cylindrical polars dVol} = r \, dr \, d\theta \, dx$$

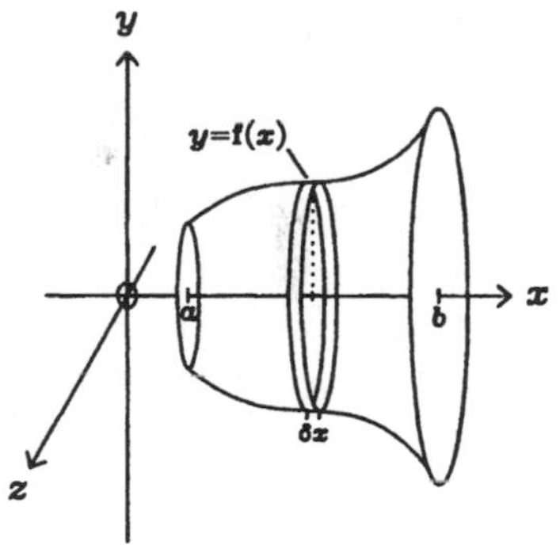

$$\therefore \text{ Volume of revolution } = \iiint_{solid} \text{dVol} = \int_{x=a}^{x=b} dx \int_{r=0}^{r=f(x)} r \, dr \int_{\theta=0}^{\theta=2\pi} d\theta$$

$$\text{But } \int_0^{2\pi} d\theta = 2\pi \quad \text{and} \quad \int_0^{f(x)} r \, dr = \left[\frac{r^2}{2}\right]_0^{f(x)} = \tfrac{1}{2}[f(x)]^2$$

$$\therefore \text{ Volume of revolution } = \underline{\pi \int_a^b y^2 \, dx}$$

Notice that, unlike in the previous example, the triple integral does not reduce to the product of three one-dimensional integrals. This is because the radius, or the maximum value of the r component, depends on the x-coordinate: $r = f(x)$. Thus, while the angular θ-integral is self-contained, the radial and lateral contributions are interlinked through the limits; indeed, the sum is very difficult to formulate if the order of the r and x integrals is reversed.

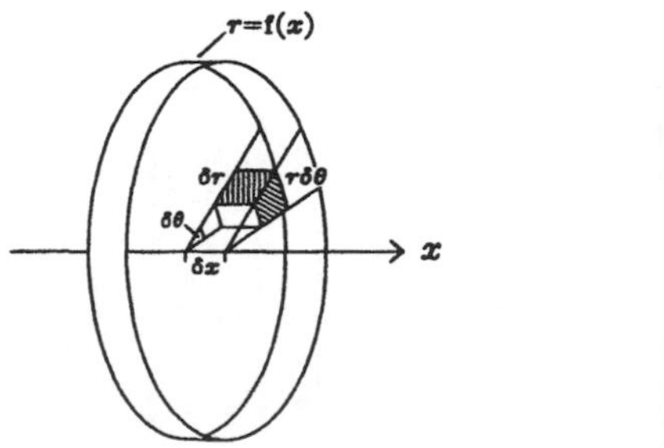

The initial evaluation of the angular and radial integrals has a simple geometrical interpretation in that it represents the volume of a thin disc at a given x-coordinate: $\pi\,[f(x)]^2\,dx$, remembering that the area of a circle is 'πR^2'. The expression for the volume of revolution can therefore be seen as the sum of many such thin discs, stacked up from $x=a$ to $x=b$.

Finally, since a sphere can be generated by rotating the semi-circular arc $x^2+y^2=R^2$ (for $y\geq 0$) through 360°, the formula of the previous example can be obtained as follows:

$$\text{Volume of sphere} = \pi\int_{-R}^{R}\left(R^2-x^2\right)\mathrm{d}x = \pi\left[R^2x-\frac{x^3}{3}\right]_{-R}^{R} = \underline{\tfrac{4}{3}\pi R^3}$$

12.4 Evaluate the double integral $\int\int x^2(1-x^2-y^2)\,\mathrm{d}x\,\mathrm{d}y$ over a circle of radius 1 centred at $x=0$ and $y=0$ in two different coordinate systems: **(a)** Cartesian, and **(b)** polar.

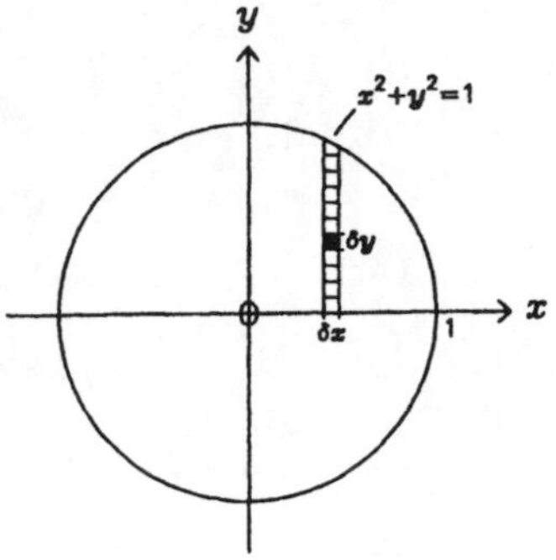

(a)
$$\iint\limits_{circle} x^2\left(1-x^2-y^2\right)\mathrm{d}x\,\mathrm{d}y = 4\int_{x=0}^{x=1}x^2\,\mathrm{d}x\int_{y=0}^{y=\sqrt{1-x^2}}\left(1-x^2-y^2\right)\mathrm{d}y$$

$$= 4\int_0^1 x^2\left[y-x^2y-\frac{y^3}{3}\right]_0^{\sqrt{1-x^2}}\mathrm{d}x$$

$$= 4\int_0^1 x^2\sqrt{1-x^2}\left[1-x^2-\frac{1-x^2}{3}\right]\mathrm{d}x$$

$$= \frac{8}{3}\int_0^1 x^2\left(1-x^2\right)^{3/2}\mathrm{d}x$$

$[x=\sin\theta]$

But
$$\int_0^1 x^2\left(1-x^2\right)^{3/2}\mathrm{d}x = \int_0^{\pi/2}\sin^2\theta\cos^4\theta\,\mathrm{d}\theta$$

and
$$\sin^2\theta\cos^4\theta = (\sin\theta\cos\theta)^2\cos^2\theta$$

$$= \frac{\sin^2 2\theta}{4}\times\frac{\cos 2\theta+1}{2}$$

$$= \frac{1-\cos 4\theta}{8}\times\frac{\cos 2\theta+1}{2}$$

$$= \frac{1+\cos 2\theta-\cos 4\theta-(\cos 2\theta+\cos 6\theta)/2}{16}$$

$$= \frac{2+\cos 2\theta-2\cos 4\theta-\cos 6\theta}{32}$$

$$\therefore \quad \iint\limits_{circle} x^2(1-x^2-y^2)\,dx\,dy = \tfrac{1}{12}\int_0^{\pi/2}(2+\cos 2\theta - 2\cos 4\theta - \cos 6\theta)\,d\theta$$

$$= \tfrac{1}{12}\left[2\theta + \frac{\sin 2\theta}{2} - \frac{\sin 4\theta}{2} - \frac{\sin 6\theta}{6}\right]_0^{\pi/2}$$

$$= \underline{\tfrac{\pi}{12}}$$

As in example 12.1, we replaced the integral over the whole circle by four times that for the positive quadrant. We are able to do this here because of the symmetry of the integrand, $x^2(1-x^2-y^2)$, which has the same value for a given magnitude of x and y independent of their signs.

(b) In polar coordinates, r and θ, the surface element $dx\,dy$ takes the form $r\,dr\,d\theta$. Therefore, putting $x = r\cos\theta$ and $y = r\sin\theta$, we have

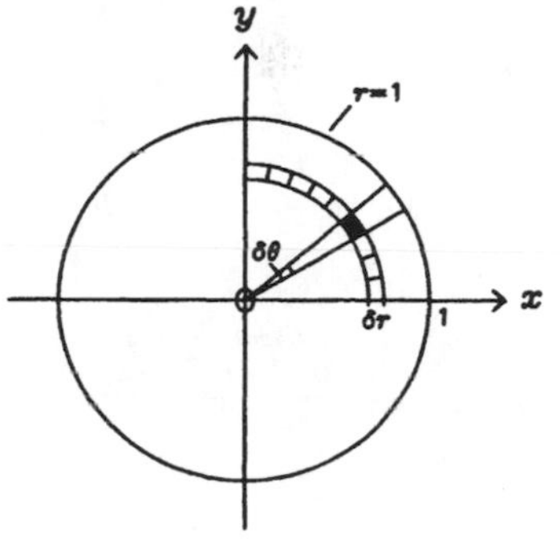

$$\iint\limits_{circle} x^2(1-x^2-y^2)\,dx\,dy = \iint\limits_{circle} r^2\cos^2\theta\,(1-r^2)\,r\,dr\,d\theta$$

$$= 4\int_{r=0}^{r=1}(r^3-r^5)\,dr\int_{\theta=0}^{\theta=\pi/2}\cos^2\theta\,d\theta$$

$$= 4\left[\frac{r^4}{4}-\frac{r^6}{6}\right]_0^1\int_0^{\pi/2}\frac{\cos 2\theta + 1}{2}\,d\theta$$

$$= 4\left(\tfrac{1}{4}-\tfrac{1}{6}\right)\left[\frac{\sin 2\theta}{4}+\frac{\theta}{2}\right]_0^{\pi/2}$$

$$= 4\times\frac{3-2}{12}\times\frac{\pi}{4} = \underline{\frac{\pi}{12}}$$

Reassuringly, we obtain the same value of the double integral using both Cartesian and polar coordinates. The effort required to do the calculation is far less for (b) than (a), however, and illustrates the point that problems are best formulated in a coordinate system which matches the geometry of the situation being considered. In this case we were integrating over a circular region, and so polar coordinates represent the most natural choice; if it had been a rectangular or triangular (with a right-angle) domain, then Cartesians would have been better.

13 Ordinary differential equations

13.1 The number N of radioactive atoms in a sample decays with time t according to the law $dN/dt = -\lambda N$. If there are originally N_0 radioactive atoms, obtain an expression for the *half-life*, the time required for the number of radioactive atoms to drop to $N_0/2$.

$$\frac{dN}{dt} = -\lambda N \qquad \Rightarrow \qquad \int_{N_0}^{N} \frac{dN}{N} = -\lambda \int_0^t dt$$

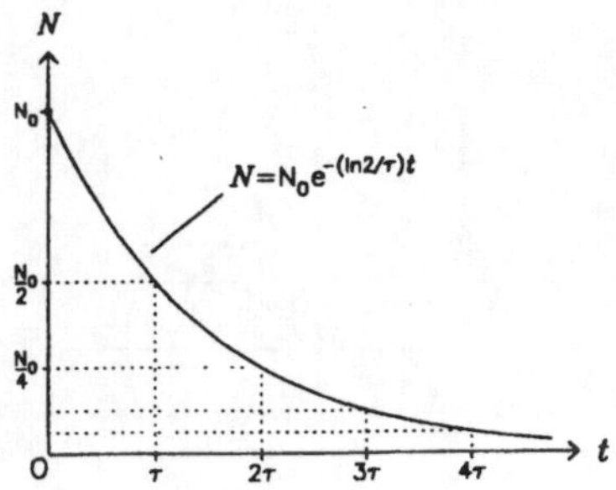

$$\therefore \quad \Big[\ln N\Big]_{N_0}^{N} = \ln\left(\frac{N}{N_0}\right) = -\lambda t \qquad \text{i.e.} \quad N = N_0\, e^{-\lambda t}$$

The *half-life* τ is defined as the time at which $N = N_0/2$.

$$\therefore \quad \frac{N_0}{2} = N_0 e^{-\lambda\tau} \quad \Leftrightarrow \quad e^{-\lambda\tau} = \tfrac{1}{2} \quad \Leftrightarrow \quad -\lambda\tau = \ln(1/2) = -\ln 2$$

$$\therefore \quad \underline{\tau = \frac{\ln 2}{\lambda} = \frac{0.693}{\lambda}}$$

13.2 Obtain general solutions of the following first-order ODEs

(a) $\dfrac{dy}{dx} = \dfrac{1-y^2}{x}$, **(b)** $\dfrac{dy}{dx} = \dfrac{2y^2+xy}{x^2}$, **(c)** $\dfrac{dy}{dx} = \dfrac{x+y+5}{x-y+2}$,

(d) $\dfrac{dy}{dx} + y\cot x = \operatorname{cosec} x$, **(e)** $\dfrac{dy}{dx} + 2xy = x$, **(f)** $\dfrac{dy}{dx} + \dfrac{y}{x} = \cos x$.

(a) $$\int \frac{dy}{1-y^2} = \int \frac{dy}{(1-y)(1+y)} = \frac{1}{2}\int \left(\frac{1}{1-y} + \frac{1}{1+y}\right) dy = \int \frac{dx}{x}$$

$$\therefore \quad \tfrac{1}{2}\left[-\ln(1-y) + \ln(1+y)\right] = \ln\sqrt{\frac{1+y}{1-y}} = \ln x + A$$

Taking exponentials and squaring $\quad \underline{\dfrac{1+y}{1-y} = Bx^2} \quad (\text{with } B = e^{2A})$

(b) A homogeneous equation; use the standard substitution $y = Vx$.

$$\frac{dy}{dx} = V + x\frac{dV}{dx} = 2\left(\frac{y^2}{x^2}\right) + \frac{xy}{x^2} = 2V^2 + V$$

$$\therefore\ x\frac{dV}{dx} = 2V^2 \Rightarrow \int\frac{dx}{x} = \int\frac{dV}{2V^2}$$

$$\therefore\ \ln x = -\frac{V^{-1}}{2} + A = -\frac{1}{2}\left(\frac{x}{y}\right) + A \qquad \text{i.e. } \underline{y = \frac{-x}{2\ln x + B}}$$

(c) Transform using the linear substitutions $u = x + a$ and $v = y + b$.

$$\frac{dy}{dx} = \frac{dv}{du} = \frac{(u-a)+(v-b)+5}{(u-a)-(v-b)+2} = \frac{u+v+5-a-b}{u-v+2-a-b} \qquad du = dx,\ dv = dy$$

$$\left.\begin{matrix} 5-a-b=0 \\ 2-a+b=0 \end{matrix}\right\} \quad \begin{matrix} a = 7/2 \\ b = 3/2 \end{matrix} \quad \Rightarrow \quad \frac{dv}{du} = \frac{u+v}{u-v}$$

We next need a further substitution $v = \Theta u$.

$$\frac{dv}{du} = \Theta + u\frac{d\Theta}{du} = \frac{1+v/u}{1-v/u} = \frac{1+\Theta}{1-\Theta}$$

$$\therefore\ u\frac{d\Theta}{du} = \frac{1+\Theta}{1-\Theta} - \Theta = \frac{1+\Theta^2}{1-\Theta}$$

$$\therefore\ \int\frac{1-\Theta}{1+\Theta^2}\,d\Theta = \int\frac{d\Theta}{1+\Theta^2} - \int\frac{\Theta\,d\Theta}{1+\Theta^2} = \int\frac{du}{u}$$

$$\text{i.e. } \tan^{-1}\Theta - \tfrac{1}{2}\ln(1+\Theta^2) = \ln u + B$$

$$\therefore\ \tan^{-1}\left(\frac{v}{u}\right) - B = \ln u + \tfrac{1}{2}\ln\left[1 + \left(\frac{v}{u}\right)^2\right] = \ln\left(u\sqrt{1+\left(\frac{v}{u}\right)^2}\right) \qquad \Theta = v/u$$

$$\text{i.e. } \underline{\tan^{-1}\left(\frac{y+3/2}{x+7/2}\right) = B + \ln\sqrt{(x+7/2)^2 + (y+3/2)^2}} \qquad \begin{matrix} u = x + 7/2 \\ v = y + 3/2 \end{matrix}$$

(d) An integrating factor is needed.

$$I(x) = \exp\left(\int \cot x\,dx\right) = \exp\left(\int\frac{\cos x}{\sin x}\,dx\right) = \exp[\ln(\sin x)] = \sin x$$

$$\therefore\ \frac{d}{dx}(y\sin x) = \sin x\,\operatorname{cosec} x = 1$$

$$\therefore\ y\sin x = x + B \qquad \text{i.e. } \underline{y = \frac{x+B}{\sin x}}$$

(e) $\mathrm{I}(x) = \exp\left(\int 2x\,\mathrm{d}x\right) = \exp(x^2) \quad \therefore \quad \frac{\mathrm{d}}{\mathrm{d}x}\left(y\,\mathrm{e}^{x^2}\right) = x\exp(x^2)$

$$\therefore \quad y\,\mathrm{e}^{x^2} = \int x\,\mathrm{e}^{x^2}\,\mathrm{d}x = \frac{\mathrm{e}^{x^2}}{2} + A \qquad \text{i.e.} \quad \underline{y = \tfrac{1}{2} + A\,\mathrm{e}^{-x^2}}$$

(f) $\mathrm{I}(x) = \exp\left(\int \frac{1}{x}\,\mathrm{d}x\right) = \mathrm{e}^{\ln x} = x \quad \therefore \quad \frac{\mathrm{d}}{\mathrm{d}x}(xy) = x\cos x$

$$\therefore \quad xy = \int x\cos x\,\mathrm{d}x = x\sin x - \int \sin x\,\mathrm{d}x = x\sin x + \cos x + A$$

$$\therefore \quad \underline{y = \sin x + \frac{A + \cos x}{x}}$$

13.3 The *Bernoulli* equation is

$$\frac{\mathrm{d}y}{\mathrm{d}x} + P(x)\,y = y^{\alpha}\,Q(x).$$

Use the substitution $v = y^{-(\alpha-1)}$ to transform it into a form soluble with an integrating factor, and solve for $y(x)$ when $P(x) = Q(x) = x$, and $\alpha = 2$.

$\frac{\mathrm{d}v}{\mathrm{d}x} = -(\alpha-1)\,y^{-\alpha}\,\frac{\mathrm{d}y}{\mathrm{d}x}$

Using the substitution $v = y^{-(\alpha-1)}$, the Bernoulli equation transforms to

$$\frac{y^{\alpha}}{-(\alpha-1)}\frac{\mathrm{d}v}{\mathrm{d}x} + P(x)\,y = y^{\alpha}\,Q(x) \quad \Leftrightarrow \quad \frac{1}{1-\alpha}\frac{\mathrm{d}v}{\mathrm{d}x} + P(x)\,v = Q(x)$$

Taking the case $P(x) = Q(x) = x$ and $\alpha = 2$, we set $v = y^{-1}$

$$\therefore \quad \frac{\mathrm{d}v}{\mathrm{d}x} - x\,v = -x, \qquad \text{for which} \quad \mathrm{I}(x) = \exp\left(-\int x\,\mathrm{d}x\right) = \mathrm{e}^{-x^2/2}$$

$$\therefore \quad \frac{\mathrm{d}}{\mathrm{d}x}\left(v\,\mathrm{e}^{-x^2/2}\right) = -x\,\mathrm{e}^{-x^2/2} \Rightarrow v\,\mathrm{e}^{-x^2/2} = -\int x\,\mathrm{e}^{-x^2/2}\,\mathrm{d}x = \mathrm{e}^{-x^2/2} + A$$

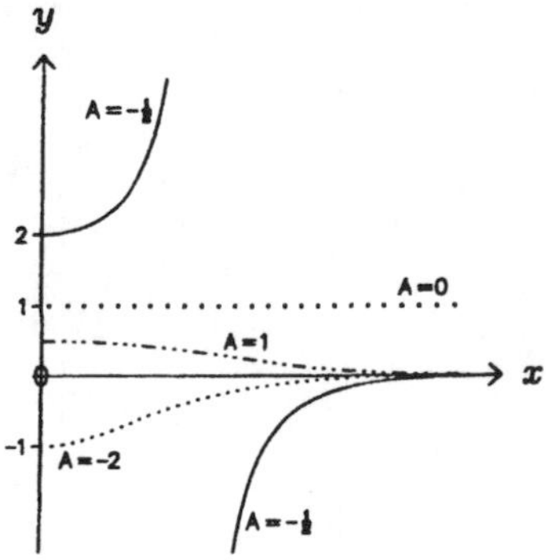

$$\therefore \quad v = 1 + A\,\mathrm{e}^{x^2/2} \quad \Leftrightarrow \quad \frac{1}{y} - 1 = A\,\mathrm{e}^{x^2/2} \qquad \text{i.e.} \quad \underline{y = \frac{1}{1 + A\,\mathrm{e}^{x^2/2}}}$$

As is often the case with non-linear ODEs, the character of the solution depends critically on the value of the integration constant A.

13.4 Show that the general linear first-order ODE can be written as $[Q(x) - P(x)\,y]\,\mathrm{d}x - \mathrm{d}y = 0$. By multiplying by $\mathrm{I}(x)$, and applying the test for exactness, show that $\mathrm{I}(x) = \exp\left(\int P(x)\,\mathrm{d}x\right)$.

$$\mathrm{I}(x)\left[\frac{\mathrm{d}y}{\mathrm{d}x} + P(x)\,y\right] = \mathrm{I}(x)\,Q(x) \quad \Leftrightarrow \quad \mathrm{I}(x)\left[Q(x) - P(x)\,y\right]\mathrm{d}x - \mathrm{I}(x)\,\mathrm{d}y = 0$$

This is only an exact differential if it satisfies the test for exactness.

$$\text{i.e.} \quad \frac{\partial}{\partial y_x}\left\{\mathrm{I}(x)\left[Q(x) - P(x)\,y\right]\right\} = \frac{\partial}{\partial x_y}\left\{-\mathrm{I}(x)\right\} \quad \Rightarrow \quad -\mathrm{I}(x)\,P(x) = -\frac{\mathrm{dI}}{\mathrm{d}x}$$

$$\therefore \quad \int \frac{\mathrm{dI}}{\mathrm{I}} = \int P(x)\,\mathrm{d}x \quad \Rightarrow \quad \ln \mathrm{I} = \int P(x)\,\mathrm{d}x \quad \text{i.e} \quad \underline{\mathrm{I}(x) = \exp\left(\int P(x)\,\mathrm{d}x\right)}$$

13.5 Find general solutions of $y'' + k_1\,y' + k_2\,y = \mathrm{F}(x)$ in the cases: **(a)** $k_1 = -2$, $k_2 = -3$, $\mathrm{F}(x) = \sin x$; **(b)** $k_1 = -2$, $k_2 = -8$, $\mathrm{F}(x) = x^2$; **(c)** $k_1 = 0$, $k_2 = \omega_0^2$, $\mathrm{F}(x) = \cos(\omega x)$; **(d)** $k_1 = 1$, $k_2 = 1$, $\mathrm{F}(x) = \cos(\omega x)$; **(e)** $k_1 = 0$, $k_2 = 4$, $\mathrm{F}(x) = \cos(2x)$. Obtain a full solution for case (a) given boundary conditions $y(0) = 0$, and y remaining finite as $x \to \infty$. What happens in case (c) when $\omega = \omega_0$? Find the steady state solution for case (d) by expressing the right-hand side as the real part of a complex exponential, and by trying $p = \mathcal{Re}\{A\exp(\mathrm{i}\omega x)\}$.

(a) Solve first for the complementary function $c(x)$: $\quad c'' - 2c' - 3c = 0$

Try $c = A\,\mathrm{e}^{mx} \;\Rightarrow\; A\,\mathrm{e}^{mx}\,(m^2 - 2m - 3) = A\,\mathrm{e}^{mx}\,(m-3)(m+1) = 0$

i.e $\;c = B\mathrm{e}^{-x} + C\mathrm{e}^{3x}$

The particular integral $p(x)$ must produce "$\sin x$" terms, so try

$$\begin{aligned} p &= D\sin x + E\cos x \\ p' &= D\cos x - E\sin x \\ p'' &= -D\sin x - E\cos x \end{aligned}$$

which, on substituting into $p'' - 2p' - 3p = \sin x$, gives

$$\begin{aligned} \sin x\,(-D + 2E - 3D) &= (1)\,\sin x \\ \cos x\,(-E - 2D - 3E) &= (0)\,\cos x \end{aligned} \qquad \left.\begin{aligned} -4D + 2E &= 1 \\ -4E - 2D &= 0 \end{aligned}\right\} \quad \begin{aligned} D &= -1/5 \\ E &= +1/10 \end{aligned}$$

i.e. $\;\underline{y = c + p = B\mathrm{e}^{-x} + C\mathrm{e}^{3x} - \tfrac{1}{5}\sin x + \tfrac{1}{10}\cos x}$

To proceed further we need two boundary conditions to fix the constants in the general solution. If y remains finite as $x \to \infty$ then $C = 0$ because otherwise $y \to \pm\infty$ as $x \to \infty$. If $y(0) = 0$ then we know by substituting $x = 0$ into the general solution that $B + 1/10 = 0$, so that $B = -1/10$.

$$\text{i.e.} \quad \underline{y = \tfrac{1}{10}\left(-\mathrm{e}^{-x} - 2\sin x + \cos x\right)}$$

(b) $c(x) = A\mathrm{e}^{mx} \;\Rightarrow\; A\mathrm{e}^{mx}(m^2 - 2m - 8) = A\mathrm{e}^{mx}(m+2)(m-4) = 0$

i.e. $c = B\mathrm{e}^{-2x} + C\mathrm{e}^{4x}$

$$\begin{aligned} p &= F + Ex + Dx^2 \\ p' &= E + 2Dx \\ p'' &= 2D \end{aligned}$$

$$\left.\begin{aligned} x^2(-8D) &= (1)\,x^2 & -8D &= 1 \\ x^1(-4D - 8E) &= (0)\,x^1 & -4D - 8E &= 0 \\ x^0(2D - 2E - 8F) &= (0)\,x^0 & 2D - 2E - 8F &= 0 \end{aligned}\right\} \begin{aligned} D &= -1/8 \\ E &= +1/16 \\ F &= -3/64 \end{aligned}$$

$$\text{i.e.} \quad \underline{y = B\mathrm{e}^{-2x} + C\mathrm{e}^{4x} - \tfrac{1}{8}x^2 + \tfrac{1}{16}x - \tfrac{3}{64}}$$

(c) $c(x) = A\mathrm{e}^{mx} \;\Rightarrow\; (m^2 + \omega_0^2) = (m + \mathrm{i}\omega_0)(m - \mathrm{i}\omega_0) = 0$

i.e. $c = A\mathrm{e}^{-\mathrm{i}\omega_0 x} + B\mathrm{e}^{\mathrm{i}\omega_0 x}$

This can easily be re-written in terms of $\cos \omega_0 x$ and $\sin \omega_0 x$.

$$\begin{aligned} c &= A(\cos \omega_0 x - \mathrm{i}\sin \omega_0 x) + B(\cos \omega_0 x + \mathrm{i}\sin \omega_0 x) \\ &= \underbrace{(A + B)}_{C}\cos \omega_0 x + \underbrace{(-\mathrm{i}A + \mathrm{i}B)}_{D}\sin \omega_0 x \end{aligned}$$

$$\begin{aligned} p &= E\cos \omega x + F\sin \omega x \\ p' &= -E\omega\sin \omega x + F\omega\cos \omega x \\ p'' &= -E\omega^2\cos \omega x - F\omega^2\sin \omega x \end{aligned}$$

$$\left.\begin{aligned} E\cos \omega x(-\omega^2 + \omega_0^2) &= (1)\cos \omega x \\ F\sin \omega x(-\omega^2 + \omega_0^2) &= (0)\sin \omega x \end{aligned}\right\} \quad \begin{aligned} E &= 1/(\omega_0^2 - \omega^2) \\ F &= 0 \end{aligned}$$

$$\text{i.e.} \quad \underline{y = C\cos \omega_0 x + D\sin \omega_0 x + \frac{\cos \omega x}{\omega_0^2 - \omega^2}}$$

Inspection of this solution shows an interesting behaviour when $\omega \to \omega_0$, y tends to infinity, or 'blows up'. This corresponds to driving a simple harmonic oscillator at its resonant frequency. Although theoretically the amplitude of the oscillation increases without limit, in real systems there comes a point where the motion no longer obeys the equation of simple harmonic motion.

(d) $c(x) = A\,e^{mx} \;\Rightarrow\; (m^2+m+1) = 0 \;\Rightarrow\; m = (-1 \pm \sqrt{1-4})/2$

i.e. $c = e^{-x/2}\left[A \sin\left(\sqrt{3}x/2\right) + B \cos\left(\sqrt{3}x/2\right)\right]$

$$\begin{aligned} p &= E \cos \omega x + F \sin \omega x \\ p' &= -E\omega \sin \omega x + F\omega \cos \omega x \\ p'' &= -E\omega^2 \cos \omega x - F\omega^2 \sin \omega x \end{aligned}$$

$$\begin{aligned} \cos \omega x(-\omega^2 E + \omega F + E) &= (1) \cos \omega x \\ \sin \omega x(-\omega^2 F - \omega E + F) &= (0) \sin \omega x \end{aligned}$$

Such equations are most conveniently solved using a matrix method.

$$\begin{pmatrix} 1-\omega^2 & \omega \\ -\omega & 1-\omega^2 \end{pmatrix} \begin{pmatrix} E \\ F \end{pmatrix} = \begin{pmatrix} 1 \\ 0 \end{pmatrix}$$

$$\begin{pmatrix} E \\ F \end{pmatrix} = \begin{pmatrix} 1-\omega^2 & \omega \\ -\omega & 1-\omega^2 \end{pmatrix}^{-1} \begin{pmatrix} 1 \\ 0 \end{pmatrix} = \frac{1}{\omega^4 - \omega^2 + 1} \begin{pmatrix} 1-\omega^2 \\ \omega \end{pmatrix}$$

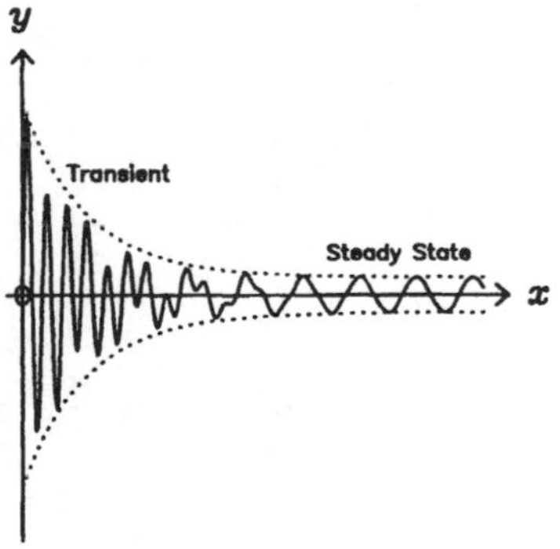

i.e. $$\underline{y = e^{-x/2}\left[A \sin\left(\sqrt{3}x/2\right) + B \cos\left(\sqrt{3}x/2\right)\right]}$$
$$\underline{+\left[(1-\omega^2) \cos \omega x + \omega \sin \omega x\right]/(\omega^4 - \omega^2 + 1)}$$

This solution describes the motion of a *driven damped oscillator*. The term on the right-hand side of the original equation represents the driving force, y is the displacement of the system and x is time. The effect of the damping (y') term is to make the complimentary function c tend to zero as $x \to \infty$; c is the *transient* solution, and p is the *steady-state* solution which describes the motion once the transient has died away.

A quicker route to this solution utilises the *complex amplitude* A

$$\begin{aligned} p &= \mathcal{R}e\{A \exp(i\omega x)\} \\ p' &= \mathcal{R}e\{A i\omega \exp(i\omega x)\} \\ p'' &= \mathcal{R}e\{A(-\omega^2) \exp(i\omega x)\} \end{aligned}$$

where the physical amplitude $|A|$, and phase ϕ are given by $A = |A|e^{i\phi}$. Substituting, and writing the right-hand side as $\mathcal{R}e\{A\exp(i\omega x)\}$, we obtain

$$(-\omega^2+i\omega+1)A\exp(i\omega x) = (1)\exp(i\omega x) \quad\Rightarrow\quad A = \frac{1}{-\omega^2+i\omega+1}$$

$$\therefore\quad \mathcal{R}e\{A\exp(i\omega x)\} = \mathcal{R}e\left\{\frac{\cos\omega x + i\sin\omega x}{-\omega^2+i\omega+1}\left[\frac{-\omega^2+1-i\omega}{-\omega^2+1-i\omega}\right]\right\}$$

$$= \underline{\frac{1}{\omega^4-\omega^2+1}\left[(1-\omega^2)\cos\omega x + \omega\sin\omega x\right]}$$

It is easy to extract the physical amplitude, in this case $1/\sqrt{\omega^4-\omega^2+1}$, and the phase, $-\arctan[\omega/(1-\omega^2)]$, from the modulus and argument of A.

(e) $c(x) = Ae^{mx} \;\Rightarrow\; (m^2+4m) = (m+2i)(m-2i) = 0 \;\Rightarrow\; m = \pm 2i$

i.e. $c = A\cos 2x + B\sin 2x$

Notice here that the function on the right-hand side of the equation is part of the complementary function, so it would be pointless to try $p = A\cos 2x + B\sin 2x$ as our first guess at the particular integral. We must instead try functions which produce "$\cos 2x$" terms when they are differentiated.

$$\begin{aligned} p &= Cx\cos 2x + Dx\sin 2x \\ p' &= C\cos 2x - 2Cx\sin 2x + D\sin 2x + 2Dx\cos 2x \\ p'' &= -2C\sin 2x - 2C\sin 2x - 4Cx\cos 2x + 2D\cos 2x + 2D\cos 2x - 4Dx\sin 2x \end{aligned}$$

$$\left.\begin{aligned} x\cos 2x(-4C+4C) &= (0)\,x\cos 2x \\ x\sin 2x(-4D+4D) &= (0)\,x\sin 2x \\ \cos 2x(4D) &= (1)\cos 2x \\ \sin 2x(-4C) &= (0)\sin 2x \end{aligned}\right\} \quad \begin{aligned} D &= 1/4 \\ C &= 0 \end{aligned}$$

i.e $\underline{y = A\cos 2x + B\sin 2x + (x\sin 2x)/4}$

We should not be too surprised that one of the coefficients (C) has turned out to be zero. The right-hand side of the original equation is said to have even *parity* because it is unchanged if x is replaced by $-x$, and the same must be true of the left-hand side. The differential operator (d^2/dx^2+4) has even parity, it too remains unchanged if $x \to -x$, and so preserves the parity of whatever it operates on. The correct p can therefore contain only even functions: $x \times \sin 2x$ is the product of two odd functions and is even, whereas $x \times \cos 2x$ is the product of an odd and an even function which has odd parity. Parity arguments are often extremely useful in quantum mechanical calculations, particularly those involving the *Schrödinger equation*.

13.6 Solve

$$x^2 \frac{d^2y}{dx^2} + 3x \frac{dy}{dx} + y = 0$$

using the trial solution $y = Ax^{\lambda}$, and then the substitution $u = \ln x$.

Try $y = Ax^{\lambda}$, so that $\frac{dy}{dx} = A\lambda x^{\lambda-1}$ and $\frac{d^2y}{dx^2} = A\lambda(\lambda-1)x^{\lambda-2}$

$$\therefore\ Ax^{\lambda}[\lambda(\lambda-1)+3\lambda+1] = 0 \quad \Rightarrow \quad \lambda(\lambda-1)+3\lambda+1 = (\lambda+1)^2 = 0$$

The repeated root produces just one solution: $\underline{y = Ax^{-1}}$

Since we know that the general solution of a second order ODE must have two arbitrary constants, we know that we have somehow missed a solution. We also know that we cannot try x times the first solution, as normally works for the $e^{\lambda x}$ solutions of the homogeneous second-order ODE, since this would produce a trial function $Bxx^{\lambda} = Bx^{\lambda+1}$ which, on substitution, produces the solution we have already found. The substitution method resolves this problem.

$$3x\frac{dy}{dx} = 3x\frac{dy}{du}\frac{du}{dx} = 3\frac{dy}{du} \quad \text{and}$$

$$x^2\frac{d^2y}{dx^2} = x^2\frac{du}{dx}\frac{d}{du}\left(\frac{1}{x}\frac{dy}{du}\right) = \frac{d^2y}{du^2} - \frac{dy}{du}$$

$$\frac{du}{dx} = \frac{1}{x}$$

$$\therefore\ \frac{d^2y}{du^2} + 2\frac{dy}{du} + y = 0$$ which we solve by substituting $y(u) = Ae^{mu}$

$$\therefore\ Ae^{mu}(m^2+2m+1) = Ae^{mu}(m+1)^2 = 0 \quad \Rightarrow \quad (m+1)^2 = 0$$

So again we get a repeated root, but now, since it is the standard homogeneous case, we know that we can write down the second solution as u times the first solution. We can then transform back using $u = \ln x$ to produce a general solution with the required two arbitrary constants:

$$y(u) = Ae^{-u} + Bue^{-u}$$

$$\Rightarrow \quad \underline{y(x) = Ae^{-\ln x} + B\ln(x)e^{-\ln x} = Ax^{-1} + B\ln(x)x^{-1}}$$

14 Partial differential equations

14.1 For the exact differential $df = y\cos(xy)\,dx + [x\cos(xy) + 2y]\,dy$, find $f(x,y)$.

If $f = f(x,y)$, then $df = \left(\frac{\partial f}{\partial x}\right)_y dx + \left(\frac{\partial f}{\partial y}\right)_x dy$

$$\therefore \text{ Exact } \Rightarrow \quad \left(\frac{\partial f}{\partial x}\right)_y = y\cos(xy) \qquad \text{— (1)}$$

$$\text{and} \quad \left(\frac{\partial f}{\partial y}\right)_x = x\cos(xy) + 2y \qquad \text{— (2)}$$

Integrating (1) with respect to x, at constant y, gives

$$f(x,y) = \sin(xy) + g(y) \qquad \text{— (3)}$$

Substituting (3) into (2), we obtain

$$\left(\frac{\partial f}{\partial y}\right)_x = x\cos(xy) + \frac{dg}{dy} = x\cos(xy) + 2y$$

$$\therefore \quad \frac{dg}{dy} = 2y \quad \Rightarrow \quad g(y) = y^2 + C$$

$$\text{i.e.} \quad \underline{f(x,y) = \sin(xy) + y^2 + C}$$

We could have done this question in a number of different ways. For example, by integrating (2) with respect to y while treating x like a constant, and substituting the resultant general formula for $f(x,y)$ into (1) to yield the related function of integration $h(x)$. Alternatively, we could even get to the solution (to within the arbitrary constant C) by directly comparing the two general formulae for $f(x,y)$ that result from the appropriate integrals of (1) and (2).

14.2 Verify that $u(x,t) = \exp(-x^2/4kt)/\sqrt{4kt}$ is a solution of the diffusion equation $\partial u/\partial t = k\,\partial^2 u/\partial x^2$.

$$\left(\frac{\partial u}{\partial t}\right)_x = \exp(-x^2/4kt)\left[\frac{\partial}{\partial t_x}\left(\frac{1}{\sqrt{4kt}}\right) + \frac{1}{\sqrt{4kt}}\frac{\partial}{\partial t_x}\left(\frac{-x^2}{4kt}\right)\right]$$

$$= \exp(-x^2/4kt)\left[-2k(4kt)^{-3/2} + \frac{x^2}{4kt^2\sqrt{4kt}}\right]$$

$$= \exp(-x^2/4kt)\left[\frac{x^2}{t} - 2k\right](4kt)^{-3/2}$$

$$\left(\frac{\partial u}{\partial x}\right)_t = \frac{\exp(-x^2/4kt)}{\sqrt{4kt}}\frac{\partial}{\partial x_t}\left(\frac{-x^2}{4kt}\right)$$

$$= -2x\exp(-x^2/4kt)\,(4kt)^{-3/2}$$

$$\therefore\quad k\frac{\partial^2 u}{\partial x^2} = k\frac{\partial}{\partial x_t}\left(\frac{\partial u}{\partial x}\right)_t$$

$$= -2k\exp(-x^2/4kt)\left[\frac{\partial}{\partial x_t}(x) + x\frac{\partial}{\partial x_t}\left(\frac{-x^2}{4kt}\right)\right](4kt)^{-3/2}$$

$$= -2k\exp(-x^2/4kt)\left[1 - \frac{x^2}{2kt}\right](4kt)^{-3/2}$$

$$\text{i.e.}\quad \underline{k\frac{\partial^2 u}{\partial x^2} = \frac{\partial u}{\partial t}}$$

This solution of the diffusion equation shows how heat (as in temperature) in a metal bar, or the concentration of a solvent in solute, spreads out with time from an initial point source. For a given time t, the spatial distribution (in x) takes the form of the ubiquitous Gaussian function (frequently met in the probability and statistics as the *'normal'* distribution). When centred at the origin, its shape is characterised by the quàdratic exponential $\exp(-x^2/2\sigma^2)$ where the width is given by the constant σ (technically known as the 'standard deviation', or its square σ^2 the 'variance'). In the diffusion case $\sigma \propto \sqrt{t}$, so that the spread doubles when the time quadruples and so on. Incidentally, the factor of $1/\sqrt{4kt}$ outside the exponential is a normalisation term which ensures that the integral of the distribution remains a constant with time; in other words, the total number of solute particles (say) is fixed even though they are spreading out. In the standard Gaussian distribution, $\exp(-x^2/2\sigma^2)$, this normalisation prefactor is $(\sigma\sqrt{2\pi})^{-1}$.

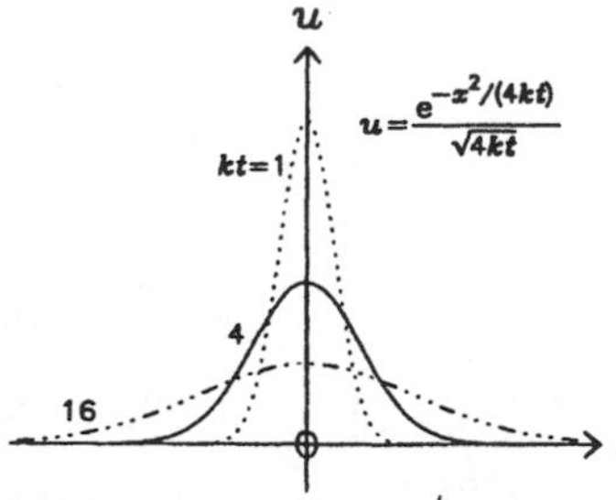

14.3 Given the Schrödinger equation for free electrons in two dimensions

$$\frac{\partial^2\psi}{\partial x^2}+\frac{\partial^2\psi}{\partial y^2}+\frac{8\pi^2 mE\,\psi}{h^2}=0$$

obtain the wavefunctions ψ, and allowed energy levels E, subject to the boundary conditions $\psi=0$ at $x=0$, $x=a$, $y=0$, and $y=b$.

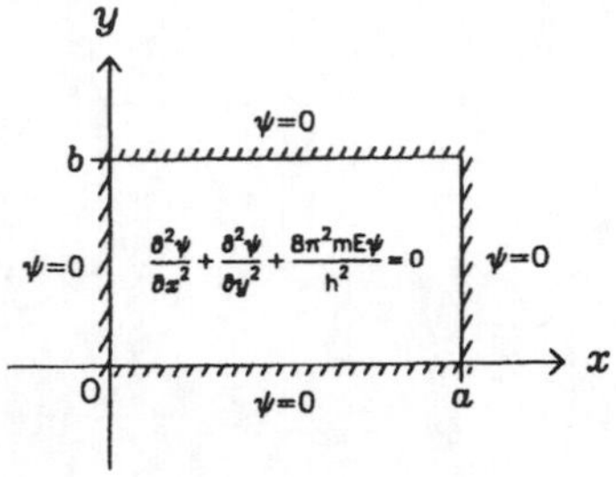

Try $\psi(x,y)=X(x)\,Y(y)$

$$\therefore\quad Y\frac{\mathrm{d}^2X}{\mathrm{d}x^2}+X\frac{\mathrm{d}^2Y}{\mathrm{d}y^2}+\frac{8\pi^2 mE\,XY}{h^2}=0$$

$$\therefore\quad \frac{1}{X}\frac{\mathrm{d}^2X}{\mathrm{d}x^2}+\frac{8\pi^2 mE}{h^2}=-\frac{1}{Y}\frac{\mathrm{d}^2Y}{\mathrm{d}y^2}=\omega^2 \qquad \text{(a constant)}$$

$$\text{i.e.}\quad \frac{\mathrm{d}^2X}{\mathrm{d}x^2}=-\left(\frac{8\pi^2 mE}{h^2}-\omega^2\right)X \quad\text{and}\quad \frac{\mathrm{d}^2Y}{\mathrm{d}y^2}=-\omega^2 Y$$

$$\therefore\quad X=A\sin(\Omega x)+B\cos(\Omega x) \quad\text{and}\quad Y=C\sin(\omega y)+D\cos(\omega y)$$

$$\text{where}\quad \Omega^2=\frac{8\pi^2 mE}{h^2}-\omega^2$$

$\therefore$ Solutions are of the form

$$\psi(x,y)=\left[A\sin(\Omega x)+B\cos(\Omega x)\right]\left[C\sin(\omega y)+D\cos(\omega y)\right]$$

But $\psi(0,y)=0 \;\Rightarrow\; B=0$

$\psi(x,0)=0 \;\Rightarrow\; D=0$

$\psi(a,y)=0 \;\Rightarrow\; \sin(\Omega a)=0$ i.e. $\Omega=k\pi/a$ for integer k

$\psi(x,b)=0 \;\Rightarrow\; \sin(\omega b)=0$ i.e. $\omega=l\pi/b$ for integer l

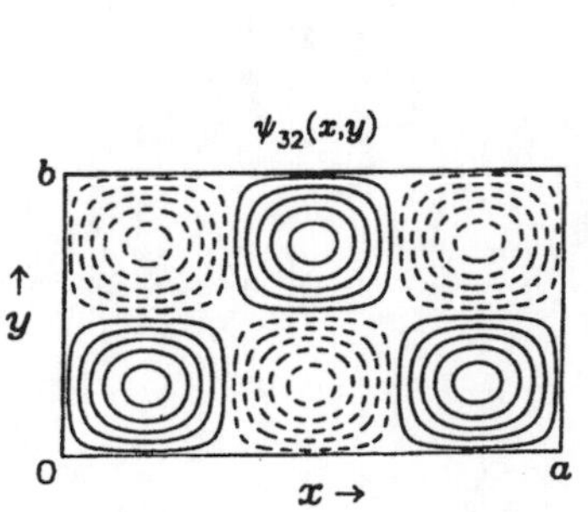

$\therefore$ Solutions are of the form $\underline{\psi_{kl}(x,y)=A_{kl}\sin(k\pi x/a)\sin(l\pi y/b)}$

where $k=1,2,3,\ldots$ and $l=1,2,3,\ldots$

But $E=\dfrac{h^2(\Omega^2+\omega^2)}{8\pi^2 m}$ $\qquad\therefore\quad \underline{E_{kl}=\dfrac{h^2}{8m}\left(\dfrac{k^2}{a^2}+\dfrac{l^2}{b^2}\right)}$

This is a two-dimensional version of the standard 'particle in the box' quantum mechanics example. In the one-dimensional case, the problem reduces graphically to one of drawing the 'normal modes' of a string whose ends are tied down and separated by a given length L (say). This leads to the condition that L can only accommodate an integer number of half-wavelengths; that is to say, $L = n\lambda/2$ where $n = 1, 2, 3, \ldots$ and λ is the wavelength. According to de Broglie, the momentum p associated with a quantum wavelength λ is given by $p = h/\lambda$ where h is Planck's constant. Thus the related kinetic energy $p^2/2m$ (where m is the mass of the 'particle') is $h^2/(2m\lambda^2) = h^2n^2/(8mL^2)$. Our formula for E_{kl} above is nothing more than the sum of two one-dimensional contributions, and $\psi_{kl}(x)$ is simply the product of the normal mode solutions in the x and y directions.

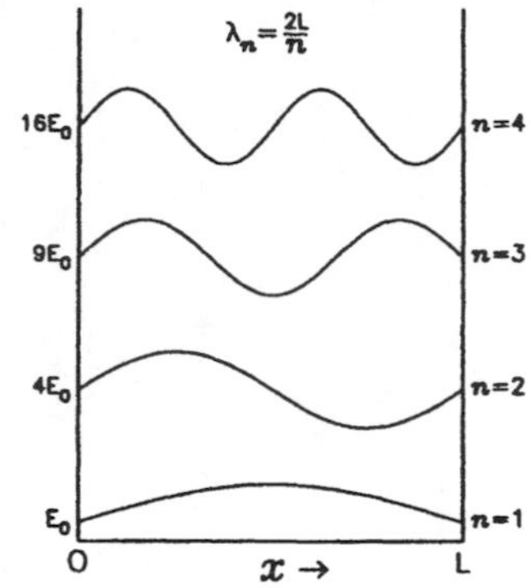

14.4 Laplace's equation in plane polar coordinates (r, θ) is

$$\frac{\partial^2 \Phi}{\partial r^2} + \frac{1}{r}\frac{\partial \Phi}{\partial r} + \frac{1}{r^2}\frac{\partial^2 \Phi}{\partial \theta^2} = 0$$

Show, by separating the variables, that there are solutions of the form

$$\Phi(r,\theta) = (A_0\theta + B_0)(C_0 \ln r + D_0) \qquad \text{and}$$

$$\Phi(r,\theta) = \left[A_p \cos(p\theta) + B_p \sin(p\theta)\right]\left(C_p r^p + D_p r^{-p}\right)$$

where the A's, B's, C's, D's and p are constants. If Φ is a single-valued function of θ, how does this restrict p? Solve the equation for $0 < r < a$ and **(a)** $\Phi(a,\theta) = T\cos\theta$, and **(b)** $\Phi(a,\theta) = T\cos^3\theta$.

Try $\quad \Phi(r,\theta) = R(r)\,\Theta(\theta)$

$$\therefore \quad \Theta\frac{d^2R}{dr^2} + \frac{\Theta}{r}\frac{dR}{dr} + \frac{R}{r^2}\frac{d^2\Theta}{d\theta^2} = 0$$

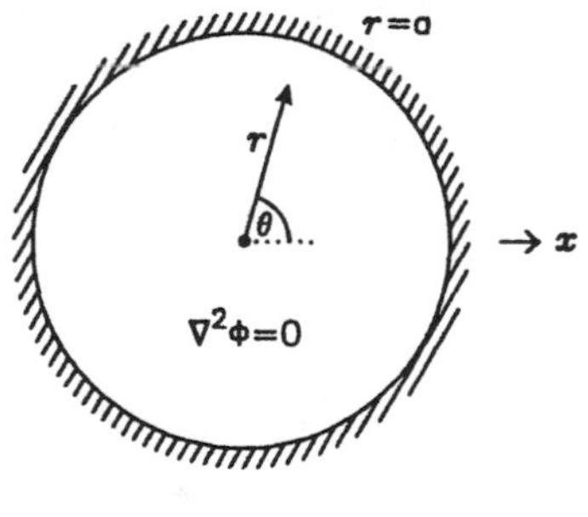

$$\therefore \quad \frac{r^2}{R}\frac{d^2R}{dr^2} + \frac{r}{R}\frac{dR}{dr} = -\frac{1}{\Theta}\frac{d^2\Theta}{d\theta^2} = p^2 \qquad \text{(a constant)}$$

$$\text{i.e.} \quad r^2\frac{d^2R}{dr^2} + r\frac{dR}{dr} = p^2R \quad \text{and} \quad \frac{d^2\Theta}{d\theta^2} = -p^2\Theta$$

Consider the special case of $p = 0$ first

$$r\frac{d^2R}{dr^2} + \frac{dR}{dr} = 0 \quad \text{and} \quad \frac{d^2\Theta}{d\theta^2} = 0$$

$$\therefore \quad \frac{\mathrm{d}}{\mathrm{d}r}\left(r\,\frac{\mathrm{d}R}{\mathrm{d}r}\right) = 0 \quad \text{and} \quad \frac{\mathrm{d}\Theta}{\mathrm{d}\theta} = A_0$$

$$\therefore \quad r\,\frac{\mathrm{d}R}{\mathrm{d}r} = C_0 \quad \text{and} \quad \Theta(\theta) = A_0\theta + B_0$$

But $\int \mathrm{d}R = C_0 \int \frac{\mathrm{d}r}{r} \quad \Rightarrow \quad R(r) = C_0 \ln r + D_0$

i.e. The $p = 0$ solution is $\underline{\Phi(r,\theta) = (A_0\theta + B_0)(C_0 \ln r + D_0)}$

More generally, if $p \neq 0$

$$\Theta = A_p \cos(p\theta) + B_p \sin(p\theta) \qquad \text{(simple harmonic motion)}$$

For $r^2\,\frac{\mathrm{d}^2R}{\mathrm{d}r^2} + r\,\frac{\mathrm{d}R}{\mathrm{d}r} = p^2 R$,

Try $R(r) = r^{\alpha}$; so that $\frac{\mathrm{d}R}{\mathrm{d}r} = \alpha r^{\alpha-1}$ and $\frac{\mathrm{d}^2R}{\mathrm{d}r^2} = \alpha(\alpha-1)\,r^{\alpha-2}$

$$\therefore \quad \alpha(\alpha-1)\,r^{\alpha} + \alpha r^{\alpha} = p^2 r^{\alpha}$$

$$\therefore \quad \alpha^2 = p^2 \quad \Rightarrow \quad \alpha = \pm p \qquad \text{i.e.} \quad R(r) = C_p r^p + D_p r^{-p}$$

i.e. The $p \neq 0$ solutions are of the form

$$\underline{\Phi(r,\theta) = \left[A_p \cos(p\theta) + B_p \sin(p\theta)\right]\left(C_p r^p + D_p r^{-p}\right)}$$

If Φ is a single-valued function of θ, then

$$\Phi(r,\theta) = \Phi(r,\theta + 2\pi n)$$

for any integer n. This cannot be satisfied for $p = 0$, but can be for the $p \neq 0$ solutions as long as p is an integer.

i.e. $\underline{p = \pm 1, \pm 2, \pm 3, \ldots}$

If the solutions are to be finite at $r = 0$, then $D_p = 0$. Therefore, the general solution is

$$\Phi(r,\theta) = \sum_p \left[A_p \cos(p\theta) + B_p \sin(p\theta)\right] C_p r^p$$

(a) $\Phi(a,\theta) = T\cos\theta$

$$\Rightarrow \quad B_p = 0 \text{ for all } p, \quad \text{and } A_p = 0 \text{ if } p \neq 1$$

$$\therefore \quad T\cos\theta = A_1 C_1 a \cos\theta \quad \Rightarrow \quad A_1 C_1 = T/a$$

i.e. The solution is $\underline{\Phi(r,\theta) = \frac{Tr}{a}\cos\theta}$

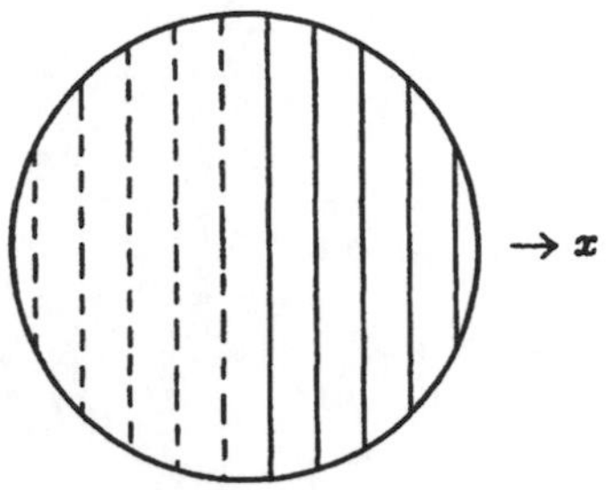

$\Phi(a,\theta)$=T$\cos\theta$

(b) $\Phi(a,\theta) = T\cos^3\theta$

$$\text{But } \cos^3\theta = \left(\frac{e^{i\theta} + e^{-i\theta}}{2}\right)^3 = \frac{e^{i3\theta} + e^{-i3\theta} + 3(e^{i\theta} + e^{-i\theta})}{8}$$

$$= \tfrac{1}{4}\left(\cos 3\theta + 3\cos\theta\right)$$

$$\therefore \quad \Phi(a,\theta) = \frac{3T}{4}\cos\theta + \frac{T}{4}\cos 3\theta$$

$$\Rightarrow \quad B_p = 0 \text{ for all } p, \quad \text{and } A_p = 0 \text{ if } p \neq 1 \text{ or } p \neq 3$$

$$\therefore \quad \Phi(r,\theta) = A_1 C_1 r \cos\theta + A_3 C_3 r^3 \cos 3\theta$$

But the boundary condition at $r = a$ implies that

$$\frac{3T}{4} = A_1 C_1 a \quad \text{and} \quad \frac{T}{4} = A_3 C_3 a^3$$

i.e. The solution is $\underline{\Phi(r,\theta) = \frac{3Tr}{4a}\cos\theta + \frac{Tr^3}{4a^3}\cos 3\theta}$

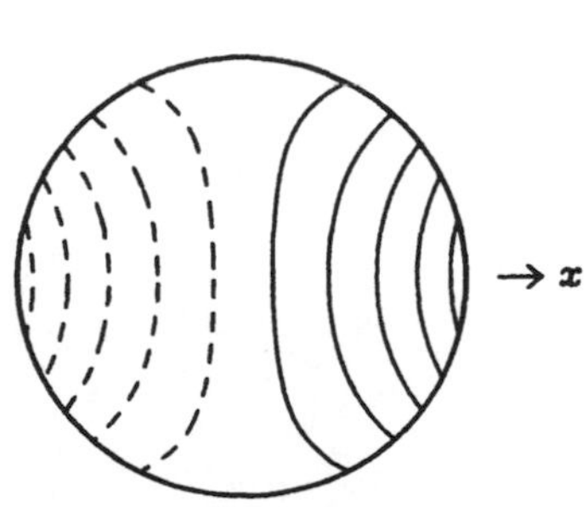

$\Phi(a,\theta)$=T$\cos^3\theta$

15 Fourier series and transforms

15.1 By first showing that the sine and cosine functions $\sin(m\omega x)$ and $\cos(n\omega x)$ are orthogonal over a period $0 \le x \le 2\pi/\omega$, derive the formulae for the coefficients of a Fourier series.

$$\int_0^{2\pi/\omega} \sin(m\omega x)\,\cos(n\omega x)\,\mathrm{d}x = \tfrac{1}{2}\int_0^{2\pi/\omega} \Big[\sin\big[(m+n)\omega x\big] + \sin\big[(m-n)\omega x\big]\Big]\,\mathrm{d}x$$

$$= -\tfrac{1}{2}\left[\frac{\cos\big[(m+n)\omega x\big]}{(m+n)\omega} + \frac{\cos\big[(m-n)\omega x\big]}{(m-n)\omega}\right]_0^{2\pi/\omega}$$

$$= \frac{1-\cos\big[2\pi(m+n)\big]}{2(m+n)\omega} + \frac{1-\cos\big[2\pi(m-n)\big]}{2(m-n)\omega}$$

$$= 0 \quad \text{for integer } m \text{ and } n$$

Strictly speaking, the above analysis assumes that $m \ne n$ (because we can't then divide by $m-n$). The result still holds for $m=n$, however, as the calculation simplifies at an earlier stage, so that $\sin\big[(m-n)\,\omega x\big]$, and all the subsequent second terms on the right, are zero.

$$\int_0^{2\pi/\omega} \sin^2(m\omega x)\,\mathrm{d}x$$

$$= \tfrac{1}{2}\int_0^{2\pi/\omega} \big[1-\cos(2m\omega x)\big]\,\mathrm{d}x$$

$$= \tfrac{1}{2}\left[x - \frac{\sin(2m\omega x)}{2m\omega}\right]_0^{2\pi/\omega}$$

$$= \tfrac{\pi}{\omega} \quad \text{for integer } m$$

$$\int_0^{2\pi/\omega} \sin(m\omega x)\,\sin(n\omega x)\,\mathrm{d}x = -\tfrac{1}{2}\int_0^{2\pi/\omega} \Big[\cos\big[(m+n)\omega x\big] - \cos\big[(m-n)\omega x\big]\Big]\,\mathrm{d}x$$

$$= -\tfrac{1}{2}\left[\frac{\sin\big[(m+n)\omega x\big]}{(m+n)\omega} - \frac{\sin\big[(m-n)\omega x\big]}{(m-n)\omega}\right]_0^{2\pi/\omega}$$

$$= \frac{\sin\big[2\pi(m-n)\big]}{2(m-n)\omega} - \frac{\sin\big[2\pi(m+n)\big]}{2(m+n)\omega}$$

$$= 0 \quad \text{if } m \text{ and } n \text{ are integers, and } m \ne n$$

$$\text{i.e.} \quad \int_0^{2\pi/\omega} \sin(m\omega x)\,\sin(n\omega x)\,\mathrm{d}x = \begin{cases} \frac{\pi}{\omega} & \text{if } m = n \\ 0 & \text{otherwise} \end{cases}$$

Similarly, $$\int_0^{2\pi/\omega} \cos(m\omega x)\cos(n\omega x)\,dx = \begin{cases} \frac{\pi}{\omega} & \text{if } m=n \\ 0 & \text{otherwise} \end{cases}$$

Therefore, $\sin(m\omega x)$ and $\cos(n\omega x)$ are orthogonal over a period $0 \le x \le 2\pi/\omega$ (for integer m and n).

Fourier series: $$f(x) = \frac{a_0}{2} + a_1\cos(\omega x) + a_2\cos(2\omega x) + a_3\cos(3\omega x) + \cdots + b_1\sin(\omega x) + b_2\sin(2\omega x) + b_3\sin(3\omega x) + \cdots$$

Multiplying both sides by $\sin(m\omega x)$ and integrating over $0 \le x \le 2\pi/\omega$, and using the orthogonality relationships just derived, we obtain

$$\int_0^{2\pi/\omega} f(x)\sin(m\omega x)\,dx = b_m\int_0^{2\pi/\omega} \sin^2(m\omega x)\,dx + 0 = \frac{\pi}{\omega}\,b_m$$

i.e. $$b_m = \frac{\omega}{\pi}\int_0^{2\pi/\omega} f(x)\sin(m\omega x)\,dx\ ; \quad \text{similarly} \quad a_m = \frac{\omega}{\pi}\int_0^{2\pi/\omega} f(x)\cos(m\omega x)\,dx$$

$$\int_0^{2\pi/\omega} f(x)\,dx = \frac{a_0}{2}\int_0^{2\pi/\omega} dx + 0 = \frac{a_0}{2}\left[x\right]_0^{2\pi/\omega} = \frac{\pi}{\omega}\,a_0$$

One of the important ideas in this example is that of orthogonal functions. While orthogonality has a natural geometrical interpretation for vectors, in that it represents perpendicular directions (having an angular separation of 90°), an equivalent physical picture for functions is not so obvious. Nevertheless, we can develop an analogy between vector manipulations and functional ones to gain a better appreciation of orthogonality in algebraic terms.

In vector analysis, two vectors $\underset{\sim}{e}_i$ and $\underset{\sim}{e}_j$, are said to be orthogonal if their scalar, or dot, product is nought when $i \neq j$

$$\underset{\sim}{e}_i \cdot \underset{\sim}{e}_j = 0 \quad \text{if } i \neq j$$

Two functions, $g_i(x)$ and $g_j(x)$, are similarly called orthogonal if the integral of their product over some specified range, $\alpha \le x \le \beta$, is zero when $i \neq j$

$$\int_\alpha^\beta g_i(x)\,g_j(x)\,dx = 0 \quad \text{if } i \neq j$$

This definition can be generalised to include a 'weighting function' $w(x)$, so that the above is written with $g_i(x)\,g_j(x)\,w(x)$ as the integrand; in the simplest, and most common, case, $w(x) = 1$.

Just as any N-dimensional vector, $\underset{\sim}{X}$, can be decomposed into a linear combination of N orthogonal basis vectors, $\underset{\sim}{e}_1, \underset{\sim}{e}_2, \underset{\sim}{e}_3, \ldots, \underset{\sim}{e}_N$

$$\underset{\sim}{X} = a_1 \underset{\sim}{e}_1 + a_2 \underset{\sim}{e}_2 + a_3 \underset{\sim}{e}_3 + \cdots + a_N \underset{\sim}{e}_N$$

so too can a function, $f(x)$, be written as a linear combination of basis functions, $g_1(x)$, $g_2(x)$, $g_3(x)$, ..., and so on.

$$f(x) = a_1 g_1(x) + a_2 g_2(x) + a_3 g_3(x) + \cdots$$

The jth coefficient, a_j, can be ascertained by taking the dot product of $\underset{\sim}{X}$ with $\underset{\sim}{e}_j$, or its functional analogue of multiplying $f(x)$ by $g_j(x)$ (and $w(x)$ if necessary) and integrating between $\alpha \le x \le \beta$

$$\underset{\sim}{X}\cdot\underset{\sim}{e}_j = a_j\,\underset{\sim}{e}_j\cdot\underset{\sim}{e}_j$$

$$a_j = \frac{\underset{\sim}{X}\cdot\underset{\sim}{e}_j}{|\underset{\sim}{e}_j|^2} \qquad \text{or} \qquad a_j = \int_\alpha^\beta f(x)\,g_j(x)\,dx \bigg/ \int_\alpha^\beta [g_j(x)]^2\,dx$$

If the basis vectors and functions are normalised, so that $|\underset{\sim}{e}_j|^2$ and $\int [g_j(x)]^2\,dx$ are unity, then these formulae can be simplified by omitting the denominators.

Given the discussion and proofs of orthogonality above, we can view a Fourier series as an expansion of $f(x)$ in terms of a set of orthogonal basis functions. This type of manipulation is very common, and useful, in theoretical studies of scientific systems, especially in quantum mechanics.

15.2 For the Fourier expansion of example 15.1, derive Parseval's identity

$$\frac{1}{\pi}\int_{-\pi}^{\pi}[f(x)]^2\,dx = \frac{a_0^2}{2} + \sum_{n=1}^{\infty}(a_n^2 + b_n^2)$$

Fourier series: ($\omega = 1$)

$$f(x) = \frac{a_0}{2} + a_1\cos x + a_2\cos 2x + a_3\cos 3x + \cdots + b_1\sin x + b_2\sin 2x + b_3\sin 3x + \cdots$$

Multiplying the left- and right-hand sides by themselves and integrating over the period $0 \le x \le 2\pi$, we find that the orthogonality of the sines and cosines reduces the integral of $[f(x)]^2$ to

$$\int_0^{2\pi}[f(x)]^2\,dx = \int_0^{2\pi}\left(\frac{a_0^2}{4} + a_1^2\cos^2 x + b_1^2\sin^2 x + a_2^2\cos^2 2x + b_2^2\sin^2 2x + \cdots\right)dx$$

$$= \frac{\pi a_0^2}{2} + \pi\left(a_1^2 + b_1^2 + a_2^2 + b_2^2 + a_3^2 + b_3^2 + \cdots\right)$$

$$\text{i.e}\quad \frac{1}{\pi}\int_0^{2\pi}[f(x)]^2\,dx = \frac{1}{\pi}\int_{-\pi}^{\pi}[f(x)]^2\,dx = \frac{a_0^2}{2} + \sum_{n=1}^{\infty}(a_n^2 + b_n^2)$$

The formula holds for an integral over $-\pi \le x \le \pi$ just as well as for $0 \le x \le 2\pi$ because our orthogonality for sines and cosines, and the implied periodicity of $f(x)$, relies on an integral over one repeat interval.

15.3 A triangular wave is represented by $f(x) = x$ for $0 < x < \pi$, $f(x) = -x$ for $-\pi < x < 0$ and $f(x) = f(x + 2m\pi)$ for any integer m. Show that its Fourier series representation is given by

$$f(x) = \frac{\pi}{2} - \frac{4}{\pi}\sum_{n=0}^{\infty}\frac{\cos[(2n+1)x]}{(2n+1)^2}$$

Period $\frac{2\pi}{\omega} = 2\pi \quad \Rightarrow \quad \omega = 1$

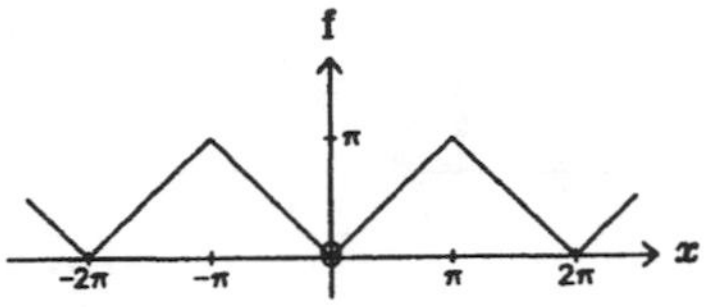

$\therefore$ *Fourier series:* $\quad f(x) = \frac{a_0}{2} + a_1\cos x + a_2\cos 2x + a_3\cos 3x + \cdots$
$\qquad\qquad + b_1\sin x + b_2\sin 2x + b_3\sin 3x + \cdots$

$$\text{where} \quad a_m = \frac{1}{\pi}\int_0^{2\pi} f(x)\cos mx\,dx$$

$$= \frac{2}{\pi}\int_0^{\pi} x\cos mx\,dx \qquad \text{(by symmetry)}$$

$$= \frac{2}{\pi}\left[x\frac{\sin mx}{m}\right]_0^{\pi} - \frac{2}{\pi m}\int_0^{\pi}\sin mx\,dx$$

$$= 0 - \frac{2}{\pi m}\left[\frac{-\cos mx}{m}\right]_0^{\pi}$$

$$= \frac{2[\cos m\pi - 1]}{\pi m^2}$$

$$= \begin{cases} 0 & \text{if } m \text{ is even, but } m \neq 0 \\ \frac{-4}{\pi m^2} & \text{if } m \text{ is odd} \end{cases}$$

$$a_0 = \frac{2}{\pi}\int_0^{\pi} x\,dx = \frac{2}{\pi}\left[\frac{x^2}{2}\right]_0^{\pi} = \pi \qquad \text{and} \qquad b_m = \frac{1}{\pi}\int_0^{2\pi} f(x)\sin mx\,dx = 0$$

(by symmetry)

$$\therefore \quad f(x) = \frac{\pi}{2} - \frac{4}{\pi}\left[\frac{\cos x}{1^2} + \frac{\cos 3x}{3^2} + \frac{\cos 5x}{5^2} + \frac{\cos 7x}{7^2} + \cdots\right]$$

i.e. *Fourier series:* $\quad \underline{f(x) = \frac{\pi}{2} - \frac{4}{\pi}\sum_{n=0}^{\infty}\frac{\cos[(2n+1)x]}{(2n+1)^2}}$

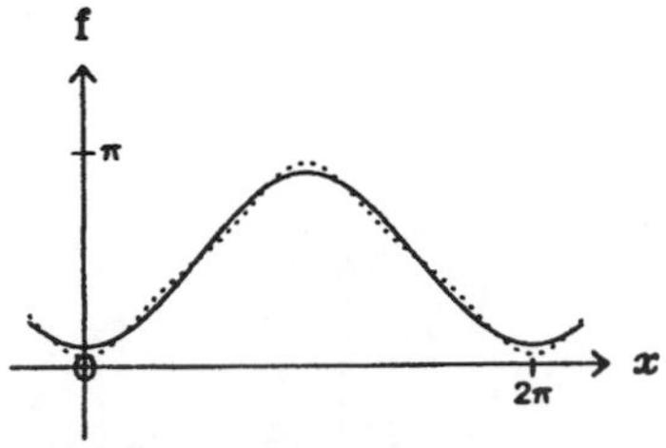

1st and 2nd-order Fourier approximations to f(x)

15.4 Derive the formulae for the intensities of the diffraction patterns from **(a)** a Young's double slit with separation d, and **(b)** a single wide slit of width D.

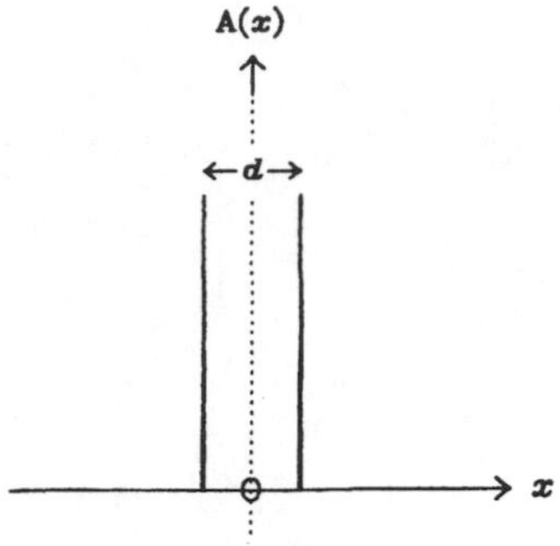

The diffraction pattern, or intensity $I(q)$, is the modulus-squared of the Fourier transform of the aperture function, $A(x)$

$$I(q) = |\psi(q)|^2 = \psi(q)^* \, \psi(q) \,, \qquad \text{where} \quad \psi(q) = \psi_0 \int_{-\infty}^{\infty} A(x) \, e^{iqx} \, dx$$

and ψ_0 is a constant (proportional to the brightness of the incident light).

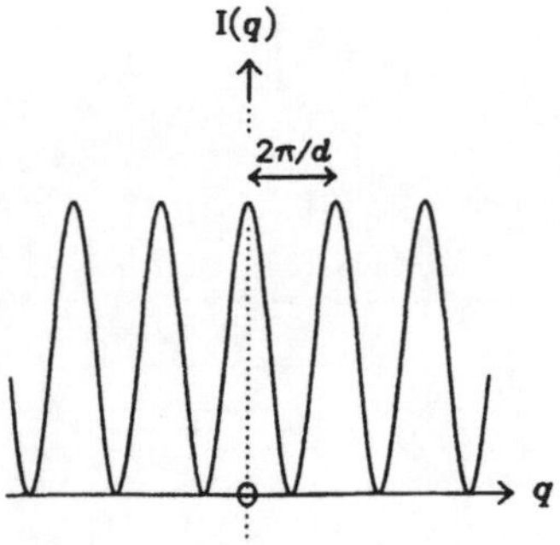

(a) For a Young's double slit, $A(x) = \delta(x+d/2) + \delta(x-d/2)$. The delta function $\delta(x-x_0)$ is an infinitely-sharp spike of unit area at $x = x_0$ (and zero otherwise), so that

$$\int_{-\infty}^{\infty} \delta(x-x_0) \, f(x) \, dx = f(x_0)$$

$$\therefore \quad \psi(q) = \psi_0 \int_{-\infty}^{\infty} \left[\delta(x+d/2) + \delta(x-d/2) \right] e^{iqx} \, dx$$

$$= \psi_0 \left[e^{-iqd/2} + e^{+iqd/2} \right]$$

$$= 2\psi_0 \cos(qd/2)$$

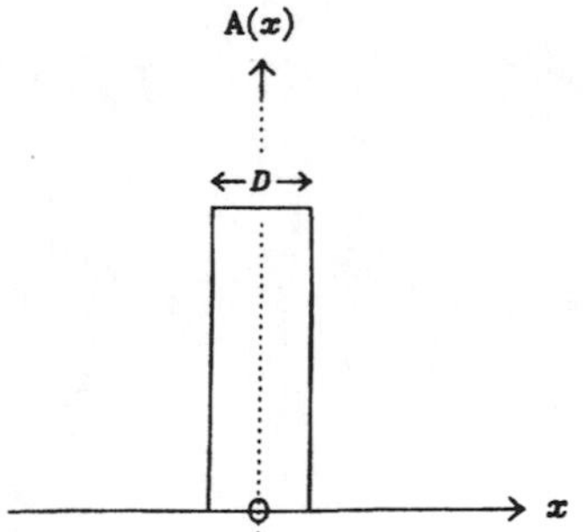

$$\therefore \quad I(q) = 4|\psi_0|^2 \cos^2(qd/2) \propto \underline{1 + \cos(qd)}$$

(b) For a wide slit, $A(x) = 1$ for $|x| < D/2$ and 0 otherwise.

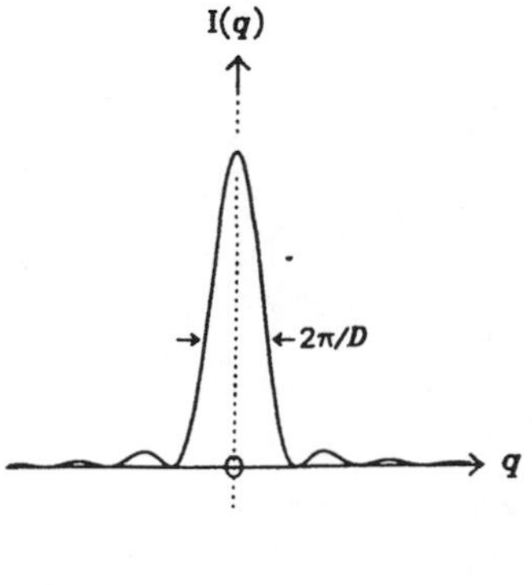

$$\therefore \quad \psi(q) = \psi_0 \int_{-D/2}^{D/2} e^{iqx} \, dx = \psi_0 \left[\frac{e^{iqx}}{iq} \right]_{-D/2}^{D/2}$$

$$= \psi_0 \, \frac{\left(e^{iqD/2} - e^{-iqD/2} \right)}{iq}$$

$$= \frac{2\psi_0}{q} \sin(qD/2)$$

$$\therefore \quad I(q) = \frac{4|\psi_0|^2}{q^2} \sin^2(qD/2) \propto \underline{\frac{1}{q^2} \left[1 - \cos(qD) \right]}$$

Other "simple" functions whose Fourier transforms are frequently met in scientific situations are:

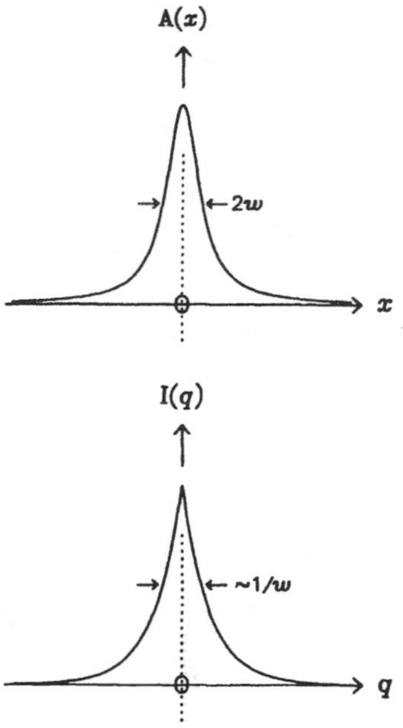

(i) a 'comb' function, or an infinite array of (equally-sized) sharp spikes with a constant separation of d between neighbouring peaks; after a bit of thought, its Fourier transform can be shown to be a comb function with a reciprocal spacing proportional to $1/d$.

(ii) a Gaussian function, as encountered in example 14.2, of the form $\exp(-x^2/\sigma^2)$, which has a width proportional to σ; its Fourier transform formally requires somewhat more advanced mathematics to derive, but is found to be a Gaussian with a reciprocal width which is of order $1/\sigma$.

(iii) a Lorentzian function, of the form $1/(x^2+w^2)$, which also falls off symmetrically from a maximum at $x=0$, but has higher tails than a Gaussian, and has a width proportional to w; the derivation of its Fourier transform also entails more advanced mathematics, but turns out to be a symmetrically decaying exponential (like $e^{-|x|}$) with a 'half-life' that scales with $1/w$.

In two, and higher, dimensions it can be useful to evaluate Fourier transforms in polar coordinates, and this tends to lead to the emergence of Bessel functions and so on.

> **15.5** Find the Fourier transform of a δ-function, or a narrow spike of unit area, located at: **(a)** the origin, $x=0$; and **(b)** $x=d$. How do they differ, and what are implications of (only) measuring the intensities?

(a) $$\psi(q) = \psi_0 \int_{-\infty}^{\infty} \delta(x)\, e^{iqx}\, dx = \underline{\psi_0}$$

(b) $$\psi(q) = \psi_0 \int_{-\infty}^{\infty} \delta(x-d)\, e^{iqx}\, dx = \underline{\psi_0\, e^{iqd}}$$

The amplitudes, or moduli, for both cases are the same, and are equal to $|\psi_0|$; the phases, or arguments, differ by a factor qd. If only the intensity $|\psi(q)|^2$ can be measured, then cases (a) and (b) will yield identical diffraction patterns. Thus, from such data, we will be able to infer that the 'object' giving rise to the diffraction pattern consisted of a single isolated spike, but we will not be able to say anything about its location.

Although we have illustrated the repercussions of the loss of phase information with a very trivial example, the problem can be a serious one for real-life scientific work.

Index